CONTENTS

Page

APPENDIX

JUNE 10, 1999 OLYMPIC PIPE LINE ACCIDENT

MONDAY, MARCH 13, 2000

U.S. SENATE,
COMMITTEE ON COMMERCE, SCIENCE, AND TRANSPORTATION,
Bellingham, WA

Hearing held pursuant to notice, at 2 p.m. at City Hall, 210 Lottie Street, Bellingham, Washington, Hon. Slade Gorton, presiding.

Staff members assigned to this hearing: Ann Begeman and Charlotte Casey, Republican Professional Staff; Carl Bentzel, Democratic Senior Counsel; and Debbie Hersman, Democratic Professional Staff.

OPENING STATEMENT OF HON. SLADE GORTON, U.S. SENATOR FROM WASHINGTON

Senator GORTON. I'm both honored and humbled to be here today with my colleague, Senator Murray, to conduct this field hearing on the Bellingham pipeline accident. This provides us with an opportunity not only to commemorate the three young citizens of Bellingham who lost their lives last June 10th, but to learn from and apply the lessons of that day to the reauthorization of the Federal Pipeline Safety Act.

This is a formal hearing of the Senate Committee on Commerce, Science, and Transportation, and the rules of the Senate and the Senate committees are much more restrictive than what most of you are accustomed to in public hearings. Only witnesses who have been invited to testify may do so. Nevertheless, the point of the hearing is to obtain information and opinions that will inform and instruct the full Commerce Committee in its work on revising the Federal law. So I invite anyone who is interested to submit written comments to the Senate Committee on Commerce, Science, and Transportation within the next 10 days. Those written comments will be made a part of the record of these proceedings. Because I'm here to listen rather than to talk, and given the length of the witness list, I hope these comments will be brief.

Until three young men were killed in a devastating liquid pipeline explosion in Bellingham last year, most of us paid little or no attention to pipeline safety. The tragic events of June 10th changed that. While pipelines continue to be the safest means of transporting liquid fuels and gas, and though accidents may be infrequent and the more than two million miles of pipelines in the United States often invisible, Bellingham has shown us that pipelines pose potential dangers that we ignore at our peril.

State government, local governments and citizen groups in this state lost no time in answering the wake-up call from Bellingham and examining what they could do to improve pipeline safety. What they found was that while there are significant actions Washington can take to prevent and respond to accidents, such as improving the state's "call before you dig" requirements, increasing public awareness and training emergency response personnel, there is a lot the state cannot do with respect to prescribing safety standards, because this area is preempted by Federal statute.

In that light, I believe that Congress has an absolute obligation substantively to revise the Federal statute. To this end I advised my colleagues on the Commerce Committee last year that I intended to be actively involved in the reauthorization process this year, and my staff and I have spent considerable time talking to and meeting with people in Washington State and with Federal regulators and industry representatives about what should be in these revisions.

Last week I cosponsored a bill, S. 2004, introduced by Senator Murray, to amend the Pipeline Safety Act. Though we still have a lot more listening to engage in, I feel the bill's fundamental direction is correct, and I hope that the hearing today will help us significantly in refining the bill.

Based on what I've heard to date, I'm committed to seeking the following changes in Federal law: First, I will support efforts to allow states greater authority to adopt and enforce safety standards for interstate pipelines, particularly in light of the absence of meaningful Federal standards. While there may be good arguments for why pipelines should be managed systemically, and why inconsistent state restrictions could erode rather than promote safety, these arguments are fatally undermined by the absence of meaningful Federal standards. To tell state and local governments as the Pipeline Safety Act effectively does, that they cannot require internal inspections of pipelines passing through their communities, under their schools and homes and senior centers when the development of Federal safety requirements is years overdue strikes me as the worse kind of Federal conceit. This increase in authority should be accompanied by an increase of grants to states to carry out pipeline safety activities.

Second, I agree with Senator Murray that we need to improve the collection and dissemination of information about pipelines to the public and the local and state officials responsible for preventing and responding to pipeline accidents. We also need to ensure that operators are collecting information necessary to assess accurately the risks of the particular line and are responding appropriately to these risks. State and local governments as well as the public should be informed about where pipelines are, what condition they are in, when they fail, and why they fail.

That said, inundating people with unwanted technical detail may lead them to ignore it entirely and may not be the best way of meeting the public's right to know. We should, however, ensure that relevant information is gathered and made available over widely accessible means like the Internet.

Third, in addition to providing an explicit mechanism for states to seek additional regulatory authority over interstate pipelines,

S. HRG. 106–1093

JUNE 10, 1999 OLYMPIC PIPE LINE ACCIDENT

FIELD HEARING

BEFORE THE

COMMITTEE ON COMMERCE, SCIENCE, AND TRANSPORTATION UNITED STATES SENATE

ONE HUNDRED SIXTH CONGRESS

SECOND SESSION

MARCH 13, 2000

Printed for the use of the Committee on Commerce, Science, and Transportation

U.S. GOVERNMENT PRINTING OFFICE

78–574 PDF WASHINGTON : 2003

For sale by the Superintendent of Documents, U.S. Government Printing Office
Internet: bookstore.gpo.gov Phone: toll free (866) 512–1800; DC area (202) 512–1800
Fax: (202) 512–2250 Mail: Stop SSOP, Washington, DC 20402–0001

SENATE COMMITTEE ON COMMERCE, SCIENCE, AND TRANSPORTATION

ONE HUNDRED SIXTH CONGRESS

SECOND SESSION

JOHN McCAIN, Arizona, *Chairman*

TED STEVENS, Alaska
CONRAD BURNS, Montana
SLADE GORTON, Washington
TRENT LOTT, Mississippi
KAY BAILEY HUTCHISON, Texas
OLYMPIA J. SNOWE, Maine
JOHN ASHCROFT, Missouri
BILL FRIST, Tennessee
SPENCER ABRAHAM, Michigan
SAM BROWNBACK, Kansas

ERNEST F. HOLLINGS, South Carolina
DANIEL K. INOUYE, Hawaii
JOHN D. ROCKEFELLER IV, West Virginia
JOHN F. KERRY, Massachusetts
JOHN B. BREAUX, Louisiana
RICHARD H. BRYAN, Nevada
BYRON L. DORGAN, North Dakota
RON WYDEN, Oregon
MAX CLELAND, Georgia

MARK BUSE, *Republican Staff Director*
MARTHA P. ALLBRIGHT, *Republican General Counsel*
KEVIN D. KAYES, *Democratic Staff Director*
MOSES BOYD, *Democratic Chief Counsel*

Federal legislation must ensure that meaningful standards for pipeline testing, monitoring and operation are adopted at the national level. Congress has directed the department of transportation to do some of this in the past, but as I mentioned before, some of the rulemakings are years overdue. To the extent that lack of funding can account for some of the delay, we should ensure additional appropriations to allow the Office of Pipeline Safety to complete the necessary rulemakings and to develop the technology needed to conduct reliable tests of pipelines.

In addition to ensuring that the Office of Pipeline Safety offers meaningful national standards, I agree with the recommendation of the Department of Transportation's Inspector General that the Office of Pipeline Safety should act upon, either to accept or to reject, the recommendations of the National Transportation Safety Board. I don't pretend to know whether the Board's recommendations that have been accumulating for years will advance safety. It's unacceptable, however, for the Office of Pipeline Safety simply to ignore them.

Fourth, I have heard from citizens' groups who support the creation of a model oversight oil spill advisory panel in Washington State. I see real value in creating such a body and imbuing it with meaningful authority not only to respond but to initiate the development of pipeline safety measures.

As I said earlier, however, the purpose of this hearing is not to lecture but to learn. Senator Murray is here with me, and I think we can both say that Congressman Metcalf and Congressman Inslee and other members of our congressional delegation would have liked to be here. The House, unfortunately, is in session this week and while we're in recess, they're in Washington, D.C. With that, I defer to Senator Murray for her opening remarks.

[The prepared statement of Senator Gorton follows:]

PREPARED STATEMENT OF HON. SLADE GORTON, U.S. SENATOR FROM WASHINGTON

I am honored to be here today with my colleague Senator Murray to conduct this field hearing on the Bellingham pipeline accident. This is an opportunity not only to commemorate the three young citizens of Bellingham who lost their lives last June 10th, but to learn from and apply the lessons of that day to the reauthorization of the federal Pipeline Safety Act. The rules of the Commerce Committee of the United States Senate, under the auspices of which this hearing is being conducted, regrettably are far more restrictive than what most of you are used to from public hearings. Only witnesses who have been invited to testify may do so. Since the point of this hearing is to obtain information and opinions that will inform the full Commerce Committee in its work on revising federal law, however, I invite anyone who is interested to submit written comments to the Committee within 10 days. Your written comments will be made a part of the record of these proceedings.

Because I am here to listen rather than talk, and given the length of the witness list, I will keep my comments brief.

Until three young men were killed in a devastating liquid pipeline explosion in Bellingham, Washington, last year, most of us paid little or no attention to pipeline safety. The tragic events of June 10, 1999, changed that. While pipelines continue to be the safest means of transporting liquid fuels and gas, and though accidents may be infrequent and the more than two million miles of pipelines in the U.S., often invisible, Bellingham has shown us that pipelines pose potential dangers that we ignore at our peril.

State government, local government, and citizen groups in this state lost no time in answering the wake-up call from Bellingham and examining what they could do to improve pipeline safety. What they found was that while there are significant actions Washington can take to prevent and respond to accidents, such as improving the state's call-before-you-dig requirements, increasing public awareness, and training emergency response personnel, there is a lot the state cannot do with respect to prescribing safety standards because this area is preempted by federal law.

In light of this, I believe that Congress has an absolute obligation substantively to revise this federal law. To this end, I advised my colleagues on the Commerce Committee last year that I intended to be actively involved in the reauthorization process this year, and my staff and I have spent considerable time talking to and meeting with people in Washington state and with federal regulators and industry representatives about what should be in these revisions. Last week I co-sponsored a bill, S. 2004, introduced by Senator Murray to amend the Pipeline Safety Act. Though we still have a lot more listening to do, I feel the bill's fundamental direction is right and I hope that the hearing today will help us significantly in refining the measure.

Based on what I have heard to date, I am committed to seeking the following changes in federal law:

First, I support efforts to allow states greater authority to adopt and enforce safety standards for interstate pipelines, particularly in light of the absence of meaningful federal standards. While there may be good arguments for why pipelines should be managed systemically and why inconsistent state prescriptions could erode rather than promote safety, these arguments are fatally undermined by the absence of meaningful federal standards. To tell state and local governments, as the Pipeline Safety Act effectively does, that they cannot require internal inspections of pipelines passing through their communities, under their schools and homes and senior centers, when the development of federal safety requirements is years overdue, strikes me as the worst kind of federal conceit. This increase in authority should be accompanied by an increase in grants to states to carry out pipeline safety activities.

Second, I agree with Senator Murray that we need to improve the collection and dissemination of information about pipelines to the public and to local and state officials responsible for preventing and responding to pipeline accidents. We also need to ensure that operators are collecting information necessary to assess accurately the risks to the particular line and are responding appropriately to these risks. State and local governments as well as the public should be informed about where pipelines are, what condition they are in, when they fail (we need to lower the threshold for reporting failures), and why they fail. That said, inundating people with unwanted technical detail may lead them to ignore it entirely and may not be the best way of meeting the public's right to know. We should, however, ensure that relevant information is gathered and made available over widely accessible means like the Internet.

Third, in addition to providing an explicit mechanism for states to seek additional regulatory authority over interstate pipelines, federal legislation must ensure that meaningful standards for pipeline testing, monitoring, and operation are adopted at the national level. Congress has directed the DOT to do some of this in the past, but as I mentioned before, some of the rulemakings are years overdue. To the extent that lack of funding can account for some of the delay we should ensure sufficient appropriations to allow OPS to complete the necessary rulemakings and develop the technology needed to conduct reliable tests of pipelines.

In addition to ensuring that OPS adopts meaningful national standards, I agree with the recommendation of the DOT's Inspector General that OPS should act upon, either to reject or accept, the recommendations of the National Transportation Safety Board. I don't pretend to know whether NTSB's recommendations, that have been accumulating for years, will advance safety. It is unacceptable, however, that OPS simply ignore them.

Fourth, I have heard from citizens' groups who support the creation of a model oversight oil spill advisory panel in Washington state. I see a real value in creating such a body, and imbuing it with meaningful authority not only to respond to but to initiate the development of pipeline safety measures.

As I said earlier, however, the purpose of this hearing is not to lecture but to learn. That said, I invite my colleague, Senator Murray's, opening remarks.

STATEMENT OF HON. PATTY MURRAY, U.S. SENATOR FROM WASHINGTON

Senator MURRAY. Thank you very much to Senator Gorton for calling this hearing and lending your leadership to this very important cause.

I also want to thank all of the panelists who have come here today who are taking their time to testify on this very important issue. I'm looking forward to hearing all of their comments.

Today's hearing is one more step in a process to make pipelines safer and certainly has been a group effort. I want to first commend you Mayor Mark Asmundson for the tremendous amount of work you have done. He's done more than anybody I know to educate the public about this issue and to call for higher safety standards, and we all thank you for your tremendous amount of work.

I also want to take this opportunity to thank Transportation Secretary Rodney Slater for his sensitivity and his quick response to me in positioning a pipeline inspector here in the State of Washington.

I also want to thank Governor Locke for convening a task force in response to this accident, and also to your representatives, Kelli Linville and Harriet Spanel, who are here today and the tremendous work that they've done in this session of the legislature to move this issue forward, but most of all, I would like to really express my appreciation and gratitude to the families of the victims who are here today. I can't imagine how difficult it must be to live with this tragedy, and I want to applaud the courage all of you have shown all of us.

I want to tell the families that are here today that I will not stop working until we have changed our nation's laws to makes it less likely that another family will experience your loss. We owe all of you at least that much.

I wish that we didn't have to be here today. I wish this community was whole again. I wish that June 10th, 1999, was just another pleasant summer day instead of a black mark in all of our memory.

I remember that day well when my sister, who lives here in Bellingham and works at Shuksan Middle School, called me within hours after the accident to tell me frantically what had occurred here. I couldn't imagine the loss that she was describing and the scenes that she was describing.

When I came here to Bellingham a few weeks later and saw what had occurred, I was just absolutely amazed. One and a half miles of creek side was reduced to ashes in an instant. A salmon spawning ground that I was actually supposed to dedicate just a few weeks after the accident was nothing but an environmental disaster area.

When I first started looking at this, I thought that the Bellingham disaster was a freak, a fluke, something that hardly ever happened. I have been amazed to find out as I've started to investigate this issue at what I have been astonished to learn.

We have a map here that shows a sampling, a sampling of some of the major pipeline accidents that have occurred in the last 20 years. It shows you how far reaching this problem is. I want to tell you some of the statistics.

Since 1986, 14 years ago, there have been 5,700 pipeline accidents, 5,700 accidents in the last 14 years. These accidents have killed 325 people and have injured another 1,500 people. They have shattered communities from coast to coast, and there are literally hundreds of Bellinghams out there that have happened and hundreds more waiting to happen.

On average there is one reported pipeline spill in our country every single day. These accidents have destroyed families like they have here in Bellingham, and they have destroyed our environment. Each year six million gallons of hazardous liquids are released. That's like having an oil spill the size of the Exxon Valdez every 2 years. This environmental damage has been estimated to cost almost a billion dollars in the last 14 years.

Now, it's true, and we all know that pipelines offer one of the safest ways to move these hazardous materials. Statistically, they are much safer than using trucks or barges, and all of us rely on the pipelines to bring us the fuel we need to heat our homes and power our cars, but none of us should accept the status quo.

Unfortunately, efforts to improve safety have not worked. Recent events tell the story. In 1997 we witnessed the third highest net loss of material since the Office of Pipeline Safety began keeping records. 1998 was the worse year for property damage, and 1999 was tied for the second worse year in fatalities. The changes that have been made so far have not worked, and we must do more. In fact, environmentalists and the National Transportation Safety Board have been complaining about safety problems and lax regulations for years. Specific recommendations from NTSB have gone unheeded and ignored for more than a decade, and I find that unacceptable. That's why a few months ago I asked the Inspector General of the Department of Transportation to investigate the policies and practices of the Office of Pipeline Safety.

On last Friday, I received the inspector general's final report. Not surprisingly they were critical of the lack of pipeline regulation and called on Congress to force the Office of Pipeline Safety to issue long overdue safety rules. The report also notes that had we need significant investments in research and development to better test and inspect our pipelines.

To date the Office of Pipeline Safety has failed to address congressional mandates in training, testing and other key areas. While I'm pleased that had they have recently committed to fulfilling our congressional requirements, I want them to know that they have to follow through on this commitment. I believe that we can and must do better, and the time to act is now.

I want to make sure that we don't just talk about making pipeline safety. We need to make pipelines safer. That's why last January after researching this issue for a number of months, I wrote and introduced a bill that will make changes to improvement the pipeline safety in this country. My bill has in-depth testimony that, Mr. Chairman, I will submit for the record, but basically it expands state authority, and improve inspection practices and prevention practices. I was shocked to find out that we only require inspection of these pipelines when they're first laid. We in our bill require them to be routinely inspected at least every 5 years, more if the geography of the region requires it. We invest in new safety technology. We have not done enough to improve the safety technology to inspect these pipelines. It expands the public's right to know. Everybody that lives or works or goes to school near these pipelines has a right to know when they were last inspected, what was found and what has been done to repair the pipes, and finally we increased the funding to improve pipeline safety. I have been working closely with Congressmen Metcalf and Inslee and other House members and along with Sentor Gorton, and I believe that we have to act this year in this session of Congress before another tragedy occurs.

In closing let me say that we cannot undue what happened here last June. We still don't know why it happened, but we can learn from it, and we can change the law so it doesn't happen again. I hope that in the coming days and weeks we can work together to put the lessons of the Bellingham tragedy into Federal law. Never again should our children be afraid to play outside. Never again should our environment be scarred by pipeline disasters, and never again should another community suffer what Bellingham has gone through this past year. Our work will only be done when families can feel confident that the pipelines near their homes are safe.

Thank you, Mr. Chairman.

[The prepared statement of Senator Murray follows:]

PREPARED STATEMENT OF HON. PATTY MURRAY, U.S. SENATOR FROM WASHINGTON

I want to thank my colleague, Senator Gorton, for calling this hearing and for lending his leadership to this important cause. Let me also thank our panelists for coming today. I'm eager to hear your comments. I'm going to take what I learn from all of you today back with me to the Senate and use it as we work to change the law.

Today's hearing is one more step in a process to make pipelines safer, and this has certainly been a group effort.

- I'd like to thank Mayor Asmundson. He has done more than anyone I know to educate the public about pipeline safety and to call for higher safety standards.
- I want to thank Transportation Secretary Rodney Slater for his sensitivity and for his quick response in stationing a pipeline inspector here in Washington state last year.
- And I'd like to thank Governor Locke for convening his task force in response to the accident.

But most of all, I'd like to show my appreciation to the families of the victims for being here today. I can't imagine how difficult it must be to live with this tragedy, and I applaud the courage you have shown us all.

I want to tell the families that I will not stop working until we have changed our nation's laws to make it less likely that another family will experience your loss. We owe you at least that much.

I wish we didn't have to be here today. I wish this community were whole again. I wish June 10, 1999 was just another pleasant summer day—instead of a black mark in our memory.

I'll never forget how I first heard about the explosion. That evening, I stepped off a plane from Washington, D.C. into Sea-Tac airport, and my cell phone started ringing almost immediately.

It was my twin sister, who lives here in Bellingham where she works as a middle school teacher.

Her voice was frantic. She said, "Patty, have you heard? Our whole world just blew up!"

At first, I didn't know what she was talking about. Then she told me that a pipeline running directly under the parking lot of her school had blown up. It was just a block away from her classroom, and it took place just hours after the last student had left.

The explosion rocked the school. Since it happened in the weeks just after Columbine, many teachers raced from their classroom fearing the worst. Instead, they encountered a nightmare of a different sort—a hailstorm of burning branches falling into their school parking lot, singeing their clothing and leaving them in fear.

I know I don't have to recount the events of that day for any of you. You experienced them and were shaped by them, and many of you have shared your own stories with me.

I came to Bellingham a short time after the accident, and I was amazed at the wreckage I saw:

- One and a half miles of creek side were reduced to ashes in an instant.
- A salmon spawning ground I was to have dedicated a few weeks later had been turned into an environmental disaster area.

At first, I thought the explosion was a fluke—something that hardly ever happened. But then I started to investigate the issue, and I was astonished by what I learned.

This map shows a sampling of some of the major pipeline accidents in the past 20 years. It gives you a sense of how far reaching the problem is.

Now I'd like to point out some statistics that show the *frequency* of pipeline accidents.

Since 1986:

- There have been more than 5,700 pipeline accidents—5,700.
- These accidents have killed 325 people and have injured another 1,500 people.
- They have shattered communities from coast to coast. There are literally hundreds of "Bellinghams" out there, and there are hundreds more waiting to happen.
- On average, there is one reported pipeline spill in our country every day.

Not only have these accidents destroyed families, they have destroyed the environment. Each year, 6 million gallons of hazardous liquid are released. That's like having an oil spill the size of the Exxon Valdez disaster every two years. This environmental damage has been estimated to cost *almost* $1 billion.

Now it is true pipelines offer the safest way to move these hazardous materials around. Statistically, they are much safer than using trucks or barges. And each of us relies on pipelines to bring us the fuel we need to heat our homes and power our cars. But none of us should accept the status quo.

Unfortunately, efforts to improve safety haven't worked. Recent events tell the story:

- 1997 witnessed the third-highest net loss of material since the Office of Pipeline Safety—or OPS—began keeping records.
- 1998 was the *worst* year for property damage.
- And 1999 was tied for the second-worst year in fatalities. The changes made so far have not worked. We must do more.

In fact, environmentalists and the National Transportation Safety Board—the NTSB—have been complaining about safety problems and lax regulation for years. Specific recommendations from NTSB have gone unheeded and ignored for more than a decade. I find that unacceptable.

That's why a few months ago I asked the Inspector General of the Department of Transportation to investigate the policies and practices of the Office of Pipeline Safety.

On Friday, I received the Inspector General's final report. Not surprisingly they were critical of the lack of pipeline regulation and called on Congress to force OPS to issue long-overdue safety rules. The report also noted we need significant investments in research and development to better test and inspect pipelines.

To date, the Office of Pipeline Safety has failed to address congressional mandates in training, testing and other key areas. While I'm pleased that recently they have committed to fulfilling these congressional requirements, I want them to know they must follow through on their commitment. I believe we can and must do better. And the time to act is *now*.

I want to make sure that we don't just *talk* about making pipelines safer. We need to actually *make* pipelines safer. That's why in January, after researching the issue for several months, I wrote and introduced a bill that will make the changes we need to improve pipeline safety.

My bill, which is number S. 2004, is called the Pipeline Safety Act of 2000. I introduced it on January 26th. I appreciate Senator Gorton's support of my bill. I'm also pleased Senators Inouye, Lautenberg, and Bayh have co-sponsored my bill as well.

To gain support for this effort, I went door-to-door and met with many of my colleagues. I told them your stories, and I showed them pictures of Bellingham's pipeline explosion. Then I showed them the statistics and counted off the number of accidents that happened in their own home states.

Few other senators knew much about pipeline safety. Those discussions showed me that for too long, pipeline dangers have been "out of sight, and out of mind."

In preparing my bill, I looked at a lot of different ideas. I also reached out to industry groups, federal oversight officials, and local officials. I designed my bill to address five problem areas, and I'd like to spend a moment to review how my bill will address these problems.

Expand State Authority

The first way to improve pipeline safety is to give states more authority to oversee pipelines. Unfortunately, states have been virtually shut out of the process when it comes to regulating interstate pipelines.

While interstate activities are the responsibility of the federal government, states should be *partners* in preventing and responding to accidents. Ideally, states should be able to test and inspect pipelines within their boundaries if they have the expertise and resources to do so.

States like Washington and Virginia have asked for this authority. Other states have received notice that their authority is being stripped from them.

My bill would establish a process that would make it much more difficult for OPS to disapprove or withdraw a state's authority. My bill would give states the ability to address any objections by OPS *before* their authority is rejected or withdrawn. So the first step in our efforts is to empower states to be partners in the safety process.

Improve Inspection and Prevention Practices

The next thing we can do to make pipelines safer is to improve pipeline testing. Many pipelines are decades old, and they haven't been inspected since they were first put into the ground. I find that unacceptable. I've talked to many companies that do a good job of testing their pipelines. Unfortunately, the industry has an inconsistent record.

We must ensure pipeline operators are properly testing their pipelines for corrosion, leaks and other problems. That's why we need strong testing and inspection standards. These should include mandatory periodic internal testing, valve monitoring, the use of reliable leak detection devices, and other preventive activities.

For this to work, the operators must be required to take specific action when they discover problems. My bill would require periodic testing at least every five years with an option of more frequent testing if required.

Certification

As we test pipelines, we should make sure the people operating and inspecting them have the skills and training they need. In other fields affecting public safety—such as aviation—we have procedures in place to ensure that the people we depend on are properly trained and qualified. My bill would require individual certification of pipeline operators.

Invest In New Safety Technology

Another way to make pipelines safer is to develop the best tools to find problems in pipelines before those problems turn into disasters. Investing in the research and development of new testing and inspection devices may well be the best thing we could do to improve safety.

The lack of good technology is surprising. I didn't know that for many pipelines there are no devices available to do the type of testing that is needed. I was also surprised to learn that hydrostatic testing can have serious side-effects, such as stressing pipes and creating wastewater that is costly to dispose of. Many of our most dangerous pipelines—natural gas lines—bend and move in ways that make it impossible for any internal inspection device to accurately detect internal corrosion.

I've been told by OPS and industry representatives that there is some progress toward new technologies to detect problems in all pipelines. That is why my bill encourages more money for research and development, and today, I call on industry to partner with OPS in developing these new technologies.

Public Right To Know

Another way we can reduce the risk of pipeline tragedies is to expand the public's right to know about pipeline hazards. Too many communities are in the dark about what is going on with the pipelines that run under their homes, by their places of work and near their schools.

My bill has a very strong "right to know" provision that would require operators to inform state, local, and neighboring residents when there are problems with a pipeline. My bill requires companies to provide summaries of testing and inspection data, and my bill makes them tell us what they are doing to correct problems. Current law provides the public with little opportunity to learn what is happening around them. Without new federal legislation, we'll continue to be left in the dark about possible hazards.

Increase Funding to Improve Pipeline Safety

A final key step we must take is to make sure we have the money to improve pipeline safety. It does no good to pass new safety rules unless we also provide the money to carry them out. My bill provides funding for new state and federal pipeline safety programs. Those are the five key areas my bill will address.

Changing the Law

The next question is: how do we take these ideas and actually put them into law?

First, in Washington, D.C. the Commerce Committee needs to debate a pipeline safety bill. Senator Gorton is on the Commerce Committee, and I look forward to working with him to ensure that a bill is marked up *this* year.

I've asked for consideration of a bill. I'm pleased that Senator Inouye from Hawaii—who is the senior Democrat on the subcommittee of jurisdiction—has called for a hearing and mark-up as well.

I've also personally asked the Secretary of Transportation and the Administration to present their proposal to the Hill. They need to do it very soon. Without meaningful federal legislation, whatever temporary measures we institute will not protect us in the long-run.

I'd also like to mention the legislation that Representatives Metcalf and Inslee have introduced in the House. Their measure is similar to my bill, and I have been asking people to support it. I hope it passes in the House.

In the end, we have learned many lessons from Bellingham and the 5,700 accidents around the country. We have a good idea of what needs to be done. We need greater state involvement, more testing, better testing devices, and we need to compel OPS to act on congressional directives and give them the tools to enforce the law.

Today must not be the last day of our work to make pipelines safer. Today must be the start. And our work will only be done when we have passed a bill that addresses these critical safety issues. Again, I thank my esteemed colleague Sen. Gorton for bringing us together today to work on this issue.

In closing, let me say we can't undo what happened in June. We still don't know why it happened. But we can learn from it and we can change the law so it doesn't happen again.

I hope that in the coming days and weeks we can work together to put the lessons of the Bellingham tragedy into federal law.

- *Never again* should our children be afraid to play outside.
- *Never again* should our environment be scarred by pipeline disasters.
- And *never again* should another community suffer what Bellingham has suffered.

Our work will only be done when families can feel confident that the pipelines near their homes are safe.

Thank you.

Senator GORTON. Thank you, Senator Murray. We will first hear from the families of the victims of last June 10th, Mary and Frank King, Edward Williams and Katherine Dalen, Marilyn Robinson and Bruce Brabec. There are seats for all six of you up here.

Mr. and Mrs. King, you may start. We—whenever you have written testimony, it is, of course, a part of the record, but in the case of each of you, we're going to let you speak as you will, understanding the very difficult nature of your coming here in public today to do this. So take your time and tell us what you believe we need to hear.

STATEMENT OF FRANK KING

Mr. KING. My name is Frank King. This is my wife, Mary, and my 10-year old son died as a result of burns over 90 percent of his body, along with Stephen Tsiorvas, and Liam Wood was overcome by gas fumes and drowned.

I want to correct a quote in the Seattle Times that was made today that I do not hold Olympic Pipeline responsible for my son's or Stephen Tsiorvas' or Liam Wood's death. Unequivocally their negligence, their gross recklessness caused my son to die and Stephen Tsiorvas and Liam Wood to die.

My family wants to thank all of Whatcom County and our friends, our relatives, the rest of the families that were involved in this, because without them we could not have gotten through this horrible, horrific ordeal.

We have received so much support from our community that it is absolutely unfathomable. We live in a nice, nice place. I wouldn't wish the grief and anguish that I felt over the last 9 months on my worst enemy. It takes 9 months for a fetus to mature in a mother's womb, and it's been 9 months now, and I wish that I had the words so that I could make you feel how we actually feel, because before this happened to me, I can honestly say I had no idea how it felt.

I want to thank Senator Murray and Senator Gorton for their efforts in going forth with this. I also want you to know that I've talked with Jay Inslee. I've talked with Jack Metcalf's office, and I appreciate their efforts, and I'm sure that we're going in the right direction. I want to make sure that this kind of accident never happens again.

You, as part of the Federal Government, have two problems. You've got an industry that regulates itself. All you have to do is look at the article that I put in your briefing there about Koch Industries. Koch Industries was fined 35 million dollars by the Department of Energy and what the DOE found out in the investigation was they found the weak spots in their pipe the easy way. They let them break, and then to boot, they found out that when they did report a spill, they underreported their spillage by as much as 90 percent. Those are standards of the industry. Fix it when it breaks and lie to the public.

Secondly, you've got an Office of Pipeline Safety that was mandated to keep the public safe. They have far wide ranging authority over the pipeline industry. We don't really need anymore legislation to legislate against the pipeline industry. We need an Office of Pipeline Safety that is going to do the job that they were mandated to do. They can't demand that these guys hydrostatically test their lines. Do you know that there aren't even any mandates that they need to internally inspect—you just said that. Senator Murray just said that. They don't even have to internally inspect their lines.

As far as I'm concerned, the 35 million dollars that's spent on the Office of Pipeline Safety is a waste of taxpayers money. Somebody needs to be put in charge of the Office of Pipeline Safety that is not in the hip pocket of the oil industry. The whole idea of pipeline safety has to be revamped. There needs to be a zero spill tolerance.

Ms. DALEN. That's right.

Mr. KING. Not we only kill two or three kids a year.

Ms. DALEN. That's right.

Mr. KING. I'm going to say in a general statement the pipeline industry as a whole is an outrage. Can we really make the pipeline industry too safe? If we had no spills that harm the environment and no deaths by the pipeline industry, are we making it too safe? Nobody should have to go through this, and I can't think of a more horrible way for three children to die, and that's what they all were—was children—than to be burned over 90 percent of their bodies.

On June 10, 1999, I want to tell you what Olympic Pipe Line's behavior was, because their behavior is an outrage. It's an outrage to the citizens of the Bellingham. It's an outrage to the citizens of Washington and it's an outrage to the Federal Government.

I asked this question last October when I testified before Congress. Why is Olympic Pipe Line still operating south of Bayview? Give me a logical reason. Congressman Frank wasn't able to do it. I've written to President Clinton. I've written to Secretary Slater of the DOT. You've gotten the faxes. Jay Inslee has gotten the faxes. Janet Reno has gotten the faxes. I am not going to relent on this thing. This company needs to be shut down. They have eight employees that have information that would help us in the why question, and you mentioned we still don't know why this accident happened, but we can learn from it. We won't learn anything until we know why this accident happened. The owners of Olympic Pipe Line, Shell, Texaco, GATX, ARCO, just since June 10th, have been able to generate 2.7 billion dollars worth of revenues operating south of here. That's an outrage. They should be called on the carpet to tell us why this accident happened.

In January 1997, Olympic Pipe Line did run pigs to internally inspect their lines. In May 1997, Olympic Pipe Line wrote the Department of Energy and said they had problems in this section that split 2 years before this accident. In July 1997, an inspector from Olympic Pipe Line named R.J. Clauson goes into the park with a backhoe to inspect that section of pipe that split, and guess what he said? "Gosh, you know, I got all the equipment here, and you know, I'm ready to dig this pipe up to look at it visually," but it was too difficult to get to it. So he wrote his office back, and he said, "I didn't inspect this section of pipe. It was too difficult to get to." The DOE wrote them back and told them go in and replace this section and tell us what happened. They never got back with DOE.

So instead what Olympic Pipe Line does is they spend forty million dollars, and they build Bayview station so that they can increase their flow through the pipe by 20 percent, and somehow they ill-design Bayview station, and they have a blocked valve upstream from Bayview station that closes uncommanded 59 times in less than 6 months, 59 times. Olympic Pipe Line says hydrostatic testing is not the way to go to inspect the pipeline, and yet they hydrostatically tested the pipeline 59 times in less than 6 months, and then on June 10th it closed for the 60th time and slammed shut for the 60th time and blew a hole in the pipe that they already knew was in disrepair. That's seven inches by 27 inches, and the rest is history.

I ask the Office of Pipeline Safety didn't Olympic Pipe Line have the obligation to find out why this blocked valve slammed shut? Yeah, Rick Felder said yeah. I said, "Why haven't you followed up on it to find out?" "Well, it was never reported to us." I said, "It wasn't reported to you?" "No." "Why?" "That's a good question," was his answer. That's a good question.

This same blocked valve closes on June 10th at 3:28 and sends a shock wave that takes 23 seconds to go back up stream over Lookout Mountain almost 21 miles, 23 seconds, and split that pipe, and Olympic Pipe Line calls ARCO at 3:45, 7 minutes after this blocked valve slams shut, and there's no room in the pipe for any fuel, because it's already full of fuel. It's already full of two million sixty-three thousand gallons of fuel, a 37-inch line.

So then supposedly Olympic Pipe Line's computers go on the fritz, and Jim Hall from the NTSB testified before Congress he hasn't been able to get their computers to act that way, but then guess what Olympic Pipe Line does? They call ARCO at 4:16 and tell ARCO to turn the pumps back on, and the pumps are running at 9,000 barrels an hour, and it was at this point in time that the river of gasoline runs down Hanna Creek and into Whatcom Falls Creek and overcomes Liam Wood by fumes in a cloud fume that's 12 to 16 feet high, and he falls in the creek unconscious, which I've never heard of a person becoming unconscious because of gas fumes, and he drowns.

ARCO (sic) finally calls ARCO at 4:32 and says, "Turn off the pumps. We got a problem." Sixteen minutes later my son who works at my dealership which is a block off of Whatcom Falls Creek, and I was not fortunate enough to be there. I was unfortunately home at the time. He said, "Dad, I stood on the showroom floor, and I watched the fumes, the gas, the fire ball come down Whatcom Falls Creek" in front of his eyes. He said, "Dad, the fire ball was a 150 feet in the air," and it just rolled down Whatcom Falls Creek about a mile, and he watched it for about a mile right in front of his eyes.

I think that everybody, every businessman along Whatcom Falls Creek was absolutely horrified and thought they were dead, because Cascade Natural Gas is right across the street.

Now, let me tell you about Olympic Pipe Line, and I'm going to keep reiterating this, why is this pipeline company still operating south of here? This company is an outrage. In May 1999 when Olympic Pipe Line was trying to get this Cross Cascades pipeline to go from Seattle, I guess, over to Spokane, Ron Branson, who is Olympic Pipe Line's supervisor of product movement testifies that the company's leak detection system is nearly fail safe and is capable of detecting the smallest of leaks in 15 minutes. This leak went undetected for an hour and 34 minutes. In that same article he's quoted as saying there may have been a time when some pipeline companies treated leaks and spills as normal and acceptable part of doing business. Olympic does not, and yet if you ask Olympic Pipe Line for any maintenance records and the DOE has confirmed this when they had this last leak in August, they don't have any maintenance records, because the pipeline industry as a whole only fixes valves and pipes when they break.

Of course, now Ron Branson is pleading the 5th Amendment along with seven other Olympic Pipe Line employees, and it's beyond belief that the Federal Government can allow these people to plead the Fifth Amendment. I'm not going to say allow them to plead the Fifth Amendment. That's not the right question. I'm appalled that the Federal Government is going to allow these people to continue to operate south of here pleading the Fifth Amendment, and I maintain that had this pipeline company been shut down completely from the start, Equilon would have sacrificed these people in a heartbeat to get their 2.7 billion dollars. The lawyers that are working for these attorneys—the attorneys that are working for these employers were hired by Olympic Pipe Line. They're working for Olympic Pipe Line, not the employees. Olympic spokeswoman, Maggie Brown, poor Maggie. Every time she opens her mouth, she puts her foot further down her throat. She's quoted in the paper on March 2, 2000, saying Olympic from the beginning has been interested in finding out exactly what happened. Really? So you tell your employees to plead the Fifth Amendment and don't give the NTSB any information that might help them conclude their investigation.

Olympic says the leading cause of spills is construction damage, outside construction damage, and yet if you look at their spills over the years, 80 percent of their spills are caused from equipment failure and operator failure. Olympic tells the public that they have voluntarily closed the 16-inch line south of Bayview. I talked to Chris Hoidal of the Office of Pipeline Safety, and Chris Hoidal says, he laughed. He said without the 16-inch pipe going through Bellingham, that 16-inch pipe is useless.

Before the City of Bellingham signed their franchise agreement, Olympic Pipe Line goes into this section of pipe after they've checked their pig runs from 1997, and they find nine other sections of pipe that they know won't withstand hydrostatic testing. So they go in, and they replace these nine sections of pipe, because they're going to have a nightmare, a public relations nightmare if they have nine sections that split under hydrostatic testing. So they replace them, and they still had one split that I believe that was within 30 yards of Shuksan Middle School—Kulshan Middle School.

Now, Olympic is suing Imco. What an outrage. We didn't do it. We didn't have any responsibility in it. Imco was in there in 1994 and hit the pipe. That was the cause, and yet they knew that that pipe was in disrepair in 1997 and didn't do anything about it. This company is an outrage. It needs to be shut down. This company needs to be shut down. We need to be asked why, and I want to applaud Mark Asmundson's efforts, he's done a lot of work, and I know spent a lot of time in talking about pipeline safety, but the "why" question, since he won't return my calls, there's nothing wrong with the City of Bellingham's mayor asking the "why" question. We need to know why this accident happened. It is the first step in learning what do we have to do. There are no other steps.

I want to thank Jim Hall from the NTSB. I got to tell you I asked a reporter this morning who she thought the busiest person in government was today, and she, without, you know, batting an eyelash, she said undoubtedly Jim Hall. This man is so responsive and so sympathetic. Last week I called him at 3:10, which is 6:10 Washington, D.C. time. He's in a meeting. I leave a message for him. He called me back at 9:30, and he can't answer my question. So he calls Bob Chipkevich, and has Bob Chipkevich call me at 10:00 his time to answer my question.

Olympic Pipe Line's gross, wanton recklessness killed my little man. He was the light of my family's life. Wade was a little man that had an uplifting spirit that touched many, many people. My son said, "Dad, you know, all he had to do was come into the dealership, and he lifted my spirits." He was our joy in life. He was a joy to many people in Bellingham.

Olympic Pipe Line needs to answer why this accident happened, and they need to be forced to do it. Thank you.

[The prepared statement of Mr. King follows:]

Prepared Statement of Frank King

My name is Frank King. My 10 year-old son, Wade, died as a result of burns over 90% of his body along with his 10 year-old playmate, Stephen Tsiorvas. Liam Wood, 18, was overcome by fumes, rendered unconscious, and drowned. Three little boys died because of the gross negligence of Olympic Pipe Line Co. No one could ever explain to you what it is like to lose a child at the height of innocence.

My family wants to thank our friends, neighbors, and citizens of Bellingham and Whatcom County. We are extremely lucky to live in this Community. Because of their support we were able to find strength that we never knew we had. My family would not wish this grief and anguish on our worst enemy. It takes nine months for a fetus to mature in a mother's womb . . . and it has been nine months since this accident happened. And my family will always miss its son, brother, nephew, cousin . . . every day of our lives. I wish that I could begin to explain our families loss in a manner that you may have some comprehension of what we are feeling so you would realize the importance that this kind of accident will never be allowed to happen again anywhere.

The Federal Government has two major problems. First, the pipeline industry has no concern for public safety. Profits always come before people. Read the attached article about Koch Industries. The DOE recently fined Koch Industries $35,000,000. What the DOE found out is that Koch Industries found weaknesses in their pipeline the easy way . . . they let it break; and when Koch Industries reported a spill . . . they under reported them by as much as 90%. This is standard behavior in the pipeline industry . . . if it breaks we fix it and don't tell the amount that really leaked. This is exactly how Olympic Pipe Line runs their operation, and it is unbelievable that they are still being allowed to operate from Bayview Station south.

Secondly, the Office of Pipeline Safety (The Federal Government's regulatory agency) is in the pipeline industry's hip pocket and consequently refuse to regulate the industry. They let the industry regulate itself, even though OPS has a $35,000,000 budget and wide ranging authority over the pipeline industry. There is no doubt that if the OPS had been doing the job that it was mandated by the Federal Government to do, we would not be here today. Wade King and Stephen Tsiorvas would be skateboarding on the front sidewalk and Liam Wood would still be fishing and probably taking some college courses today.

The whole idea of pipeline safety needs to be revamped. The OPS must do the job it was mandated to do and that is to protect the public and the environment. No spills. Someone needs to be put in charge who isn't in the hip pocket of the pipeline industry. Someone needs to be put in charge of the OPS who will make the pipeline industry accountable. Put someone in charge that will welcome help from the states that want to help with oversight and regulation. And if one state finds that a regulation is good for its state, make it a regulation for all states. That way there will be consistency in regulations throughout the states. The pipeline industry's complaints really stem from the fact that they are against any kind of regulation. Can the pipeline industry ever be too safe? The OPS needs to be completely revamped so that it does the job that it was mandated to do, and accepts the help of the states that want to help. And Olympic Pipe Line Co. needs to be shut down until we know why this accident happened.

Look at the June 10, 1999 accident and Olympic Pipe Line's behavior and response. It is behavior that is an outrage . . . an outrage to the families of the victims; an outrage to the citizens of Bellingham; an outrage to the citizens of Washington State; and particularly an outrage to the people who live along this pipeline that stretches from Cherry Point to Portland. It is an outrage to the OPS and the Federal Government that this company has been allowed to continue to operate south of Bayview Station, since they refuse to help the NTSB in finding answers as to why this accident happened. Why are they still operating their pipeline? I asked this question to Congress in October. No one answered. But Olympic Pipe Line and its owners, Shell, Texaco, ARCO, and GATX, have generated in excess of $2.7 billion in revenues just since this accident happened. I have written letters to President Clinton, Vice President Gore, DOT Secretary Rodney Slayter, Kelly Coyner, head of OPS, etc. You have gotten copies of all correspondence. So far I have received no response. Olympic Pipe Line must be shut down and their employees forced to cooperate with the NTSB to determine the exact cause of this accident. It is the first step in assuring we are going in the right direction.

In January of 1997 Olympic Pipe Line ran smart pigs through their pipe. They found three anomalies (problem areas) in the very section of pipe that split on June 10, 1999. In May of 1997 Olympic Pipe Line wrote the DOE, yes the DOE, not the OPS, that they had problems in that section that split and would further evaluate those problems. The DOE wrote them back and advised Olympic Pipe Line to get back to the DOE concerning that section of pipe. Olympic Pipe Line never responded concerning this problem again. And note that this section of pipe that they had said was problematic was located in an area where there had been known construction. And Olympic Pipe Line has said on many occasions that the leading cause of pipeline spills is construction damage. But they never bothered to visually inspect this section of pipe. Instead, Olympic Pipe Line spent $40 million building Bayview Station in Burlington, which was built to allow Olympic Pipe Line the ability to increase its product flow by 20%. Bayview Station was completed in late December 1998. Between then and June 10, 1999, a block valve just north of Bayview Station closed 59 times uncommanded by operators of the pipe. 59 times it slammed shut in less than six months, virtually hydro-statically testing the section that runs through Bellingham each and every time. And what does Olympic Pipe Line have to say about hydro-static testing? On numerous occasions they have stated that it over stresses the pipe. So by their own words, they over stressed a pipe that they knew was in disrepair. That's real preventive maintenance. Obviously, I'm being facetious.

Then on June 10, 1999 this same block valve slams shut for the 60th time at 3:28PM. It sends a shock wave back up north that takes 23 seconds to split the pipe in Bellingham. It creates a hole in the pipe that is 7″ by 27″ and a crater that is 25 feet in diameter at the top and 5 feet at the bottom. It is about 10 feet deep and will hold about 60,000 gallons of gasoline. Finally, Olympic calls ARCO and tells them to turn the pumps off at 3:35PM. The pumps have been running for 7 minutes and there is no room for any gasoline in the pipe because it is full of about 2,063,000 gallons of gasoline already. Now the pipe sits there and leaks until 4:16PM. At this point in time, Olympic Pipe Line calls ARCO and tells them to turn the pumps back on. The pumps run at 9000 barrels per hour or 378,000 gallons per hour. Finally, Olympic Pipe Line calls ARCO and tells them to turn the pumps off at 4:32PM—16 minutes later. A virtual river of gasoline now pours down Hannah Creek and into Whatcom Falls Creek, creating a fume bank that is 12 to 16 feet high. It leaks for another ½ hour until 5:02PM. The resulting gas fumes ignite and explode in a fireball that runs down Whatcom Falls Creek 150 feet in the air for approximately 1½ miles and approximately ¾ mile back up Hannah Creek ending at the ruptured pipe.

Now the outrage—

- In May of 1999 Ron Berenston, Olympic's Supervisor of product movement, testifies that the Company's leak detection system is nearly fail-safe and is capable of detecting the smallest of leaks within 15 minutes. This leak went undetected for one hour and 34 minutes until it exploded in a fireball.
- In July of 1999 Olympic tells the news media that there was some sort of mastic on the pipe that split, indicating that someone had damaged the pipe, repaired it and covered it back up. Allan Beshore the NTSB's investigator told me that there was no foreign substance on the pipe that split. He further indicated that he had called Olympic Officials and criticized them for reporting this information to the press.
- Shortly after the accident 8 Olympic employees who were in the control room on June 10, 1999 invoked their Fifth Amendment rights, including Mr. Berenston. However, everyone goes back to work at their same jobs on June 11, 1999 as though nothing has happened.

- Olympic Spokesperson, Maggie Brown, is quoted in the Bellingham Herald on March 2, 2000 saying, "Olympic, from the beginning, has been interested in finding out exactly what happened." Really . . . that must be why their employees won't cooperate with NTSB. And let me add that there is no assurance that these same employees won't plead the Fifth Amendment once the criminal investigation is completed.
- Olympic says that the leading cause of spills is outside construction damage. However, if you examine the causes of their spills you will find that 80% of their spills are caused by operator error and equipment failure.
- Mr. Berenston also testified that "there may have been a time when some pipeline companies treated leaks and spills as a normal and acceptable part of doing business. Olympic does not." And yet Olympic cannot produce any type of maintenance records that shows anything was ever repaired unless there was a spill. That is preventive maintenance. Same old story, fix it when it breaks.
- Olympic says they have voluntarily closed the 16″ pipe south of Bayview Station. However, Chris Hoidal, OPS' Western Region Manager, stated that the 16″ pipe south of Bayview is useless unless the northern 16″ pipe that goes through Bellingham is running product.
- While the 37-mile stretch of pipe that runs through Bellingham lay idle, Olympic went back to its pig runs from January 1997 and found 9 sections of pipe that they knew would not withstand hydro-static testing. So before they had to comply with OPS Corrective Action Order to hydro-static test this pipe they replaced these nine sections of pipe. And they still had one section of pipe that split near Kulshan Middle School. What do you suppose the public would have thought had 10 sections split throughout this 37-mile stretch? For the first time in its 35 year history, Olympic Pipe Line Co. did some preventive maintenance.
- Olympic Pipe Line originally said that 277,000 gallons of fuel leaked out of their pipe, then revised it downward to 229,000 gallons of fuel which the news media now uses, but the OPS has not accepted. All you have to do is some simple math and you realize that the minimum amount of fuel that would have leaked out of that pipe was 420,000 gallons of gas. By the time the experts are done determining exactly how much leaked, you will find that between 650,000 and 950,000 gallons of gasoline actually leaked out of that pipe on June 10, 1999.
- Olympic Pipe Line did pig runs in January 1997, and admitted to the DOE that they had some problems in that section of pipe that split. Instead of checking that section, they spent $40 million building Bayview Station so they could increase their flow 20%. They misdesigned Bayview Station and had a block valve that closed 59 times in less than 6 months, over stressing a pipeline that is already in disrepair.
- Olympic Pipe Line goes to court and sues IMCo Construction, a local contractor, for damaging the pipe. They are suing IMCo for any costs that have been incurred trebled. IMCo was a contractor to the City of Bellingham. Is the City of Bellingham next to be sued?

Olympic Pipe Line has lied to the public before this accident happened and continues to lie to the public in the aftermath of this tragic accident. If the Pipeline Industry is to learn anything from this tragedy, it is to do exactly the opposite of what Olympic Pipe Line has done. Tell the truth. Cooperate with investigators to find out the reason or reasons why an accident has happened. Again, why is this company allowed to operate south of here without providing any answers. Shut these people down and force those employees to talk to NTSB investigators.

As we speak, there is a petition filtering through Bellingham and Whatcom County to force our Mayor not to renew Olympic Pipe Line's Franchise agreement that comes due in May, until the people of Bellingham have an answer as to why this accident happened. The people of Bellingham are aware that this accident could be duplicated again if the valve closure problem is not resolved. And they are not about to let that happen. The actions of our Mayor in recent months, make me wonder what agenda he is working. His signature was on the first Safe Bellingham petition last July. But he has neglected to ask the why question since he signed the franchise agreement with Olympic in September . . . against public sentiment.

The Chairman of the NTSB, Jim Hall (who by the way is one of the most compassionate and responsive individuals I have met throughout the last nine months), has told me that it will be at least June or September before he concludes his investigation. And even then he has admitted that the real causes of this accident may never be known unless those operators in the control room on June 10, 1999 begin to talk about what happened that day. Is pleading the Fifth Amendment an admission of guilt? As a law abiding citizen I believe it is. Shut Olympic Pipe Line down until they have told us why this accident happened and the NTSB has concurred.

It is a sad state of affairs in this country when the Justice Department goes after Microsoft as a monopoly, merely because a judge took a personal disliking to Bill Gates. Then turns right around and allows Exxon and Mobil to merge forming the largest oil company in the world. Then turns right around and tells ARCO and BP Amoco they can't merge. And then allows Olympic Pipe Line Co. to continue its operation as though nothing has happened.

Olympic Pipe Line's gross, wanton recklessness killed a "little man" that was the light of my family's life. Wade was a "little man" that had an uplifting spirit, who touched many, many people he came into contact with. He had a knack for uplifting the spirits of everyone he came in contact with. He was our joy in life. He was a joy to many, many in Bellingham. And, unfortunately, we have been forced to live without that wonderful little being for the rest of our lives. Our family misses him every day. Olympic Pipe Line needs to be shut down until they have told us why this accident happened.

Oil leaker fined $35 million

January 14, 2000, The Bellingham Herald

Washington, D.C.—Koch Industries, an oil pipeline company, found the weak points in its pipes the easy way, the government said Thursday. It simply waited for the pipelines to break and spill oil, part of a pattern of negligence that resulted Thursday in the biggest civil fine levied to a company for environmental violations.

The Environmental Protection Agency said that the company, based in Wichita, Kan., had agreed to pay a $30 million penalty for more than 300 separate spills of crude oil, gasoline and other oil products between 1990 and 1997.

Federal officials said it had taken them months to put the case together, partly because the company did business under many different names, and even when it reported spills it sometimes understated their volume by as much as 90 percent. Even after the Justice Department brought the case, Koch refused to say what pipelines it owned, officials said.

To save money, the company did not inspect its pipelines for corrosion, or pressure-test them, investigators said. "It was cheaper not to maintain them," said Steve Herman, the EPA's assistant administrator for enforcement.

And so, officials said, Koch just waited for the pipelines to fail.

The company leaked 3 million gallons of crude oil and other substances into ponds, lakes and rivers in Texas, Oklahoma, Kansas, Louisiana, Missouri and Alabama, according to Carol Browner, the administrator of the EPA, in announcing the settlement. The settlement was filed in U.S. District Court in Houston Thursday to settle charges brought by the government between 1995 and 1997.

As part of the settlement, the company will also pay $5 million to buy environmentally sensitive land and protect it from development.

"Today's landmark fine against Koch Industries for egregious violations of the Clean Water Act sends a strong message that those who try to profit by polluting our environment pay a price," Browner said.

The company offered a totally different interpretation, asserting through a high executive that Koch had reduced its pipeline leaks by 90 percent in the last decade.

Federal prosecutors file motion to test segment of pipeline

March 2, 2000, The Bellingham Herald and the Associated Press

Federal prosecutors have asked a judge to approve tests to determine what caused a pipeline rupture and resulting explosion that killed two 10-year-old boys and an 18-year-old man in Bellingham last June.

Prosecutors filed a sealed motion last week with U.S. District Judge Robert Lasnik, The Seattle Times reported Wednesday, citing unidentified sources familiar with the case.

The motion seeks permission for the National Transportation Safety Board to cut into a 20-foot section of pipe that was excavated from the rupture site.

Olympic officials said they want the pipeline tests done.

"We are as anxious as anybody to allow those tests to go forward," company spokeswoman Maggie Brown said following a Wednesday night meeting in Bellingham. "Olympic, from the beginning, has been interested in finding out exactly what happened."

The NTSB had planned to conduct the test last fall, but was halted by the U.S. Attorney's Office in Seattle, which feared the test would allow lawyers for pipeline operator Olympic Pipe Line Co. to argue that key evidence had been destroyed.

The new motion seeks to allow the tests, in addition to protecting federal prosecutors from any claims relating to altering of evidence if criminal charges are filed in the rupture and explosion, the Times cited criminal experts as saying.

The testing could tell investigators whether defects or corrosion existed before the pipeline rupture, NTSB spokesman Keith Holloway said.

"It's a key component to the investigation because it can tell us what happened," he said.

Prosecutors filing the motion notified lawyers representing Olympic, its top officials and a Bellingham construction company that previously dug near the pipeline, the Times reported. Investigators have looked into whether the pipe had been damaged by excavation work, and Olympic has filed a motion blaming the excavation contractor for the rupture.

A federal grand jury is investigating whether environmental or pipeline-safety laws were violated. No charges have been filed, and the investigation is expected to continue for months.

The families of the two boys killed have filed a lawsuit against Olympic.

Olympic could face a fine of as much as $10,000 in the Nov. 17 incident, when a significant amount of fuel was leaked during cleanup of the Bellingham spill, resulting in some environmental damage, state Department of Ecology spokesman Ron Langley said.

The company was told Friday it had violated state water-quality laws and now has 30 days to respond, he said.

Olympic had failed to clean silt from a perforated pipe that draws fuel and contaminated groundwater from soils in Whatcom Falls Park, causing an overflow.

Olympic spokeswoman Maggie Brown said the amount of fuel spilled was too small to be measured.

"It was a sheen," she said.

But Langley said the cleanup system leaked for 30 hours before a state inspector noticed gas in the creek.

Olympic has appealed a $120,000 fine from Ecology for the June 10 spill of 229,000 gallons of petroleum fuel.

Senate Hearing March 13, 2000

Additional Testimony From Frank King

The question that was asked Senators Gorton and Murray still remains unanswered. Why is Olympic Pipe Line allowed to continue to operate their pipeline south of Bayview Station, when their employees refuse to help the NTSB investigate this accident? Senator Gorton asked that question of Mr. Gast from Olympic Pipe Line and Mr. Felder from the Office of Pipeline Safety. Both responded by saying that they felt the 20" line south of Bayview Station was safe and operating at 80% of normal operating pressure. My question does not ask if the pipeline is safe!!! This Company needs to be shut down, until such time their employees help the NTSB put together all the answers to the why question.

Olympic Pipe Line did internal pig runs on their pipeline in January of 1997. In May of 1997, they wrote the Washington State Department of Ecology and advised them that they had three serious anomalies in that very section of pipe that split on June 10, 1999. The DOE wrote them back and told Olympic pipeline to go into the area and dig up that very section of pipe and replace it or repair it as necessary. In July of 1997, R. J. Klasen, Field Supervisor for Olympic Pipe Line, goes into the area with a crew and a backhoe to dig up that very section of pipe. He then reports back to his superiors at Olympic Pipe Line and the DOE that he did not dig up that very section of pipe because it was "too difficult to get to." Olympic Pipe Line's gross, wanton, reckless negligence murdered my son, Wade, as well as Stephen Tsiorvas and Liam Wood. This same gross, wanton, reckless negligence is the same reason that these people need to be shut down until we know why this accident happened. Shut them down now.

As I sat in that Senate Hearing room and listened to Mr. Felder's testimony and heard him say that the OPS has gone back to all the pipeline operators and asked them all to come up with safety action plans for their pipelines, I couldn't help but become extremely uneasy. That is exactly the type of attitude that must be eliminated out of the OPS. I thought that was why the OPS was created . . . to regulate the safety of the pipeline industry . . . not to ask the pipeline industry how it wanted to be regulated. Go back and list all the recommendations that the NTSB has given to the OPS over the last 30 years. List them all, and then make those the regulations by which the pipeline industry is regulated. Then add one regulation. When a pipeline company has an accident, its entire operations will be suspended until the cause of the accident has been determined. It was very apparent that the NTSB has seen very little change in the OPS' attitude to force the pipeline industry to safeguard the public. The OPS needs to be a stronger advocate for public safety and needs to welcome states help to mandate that public safety.

This morning I read in the Bellingham Herald that a maintenance supervisor at Alaska Airlines had been placed on administrative leave because he had bullied the mechanics who maintain the airplanes into not doing the proper job in their maintenance efforts. Olympic Pipe Line Co. needs to be put on administrative leave, until the NTSB has found the answers as to why this accident happened.

Senator GORTON. Thank you, Mr. King.

Does Mrs. King wish to add anything?

Mr. KING. Pardon?

Senator GORTON. Does Mrs. King wish to add anything?

Mrs. KING. No.

Senator GORTON. Ms. Dalen, we'll go to you.

STATEMENT OF KATHERINE DALEN

Ms. DALEN. I'm a little nervous about squeals. The young man that was sitting next to me is not Mr. Williams who has been busy working with teachers down around the Seattle/Tacoma area today. This man is my son and Stephen's brother, just turned 18, so I hope you don't mind. He's been kind of supportive.

Senator GORTON. Fine. Fine. You go ahead.

Ms. DALEN. I wish to thank Senator Gorton, Senator Murray and other distinguished members of the Committee on Commerce, Science, and Transportation for inviting me to speak today.

My name is Katherine Dalen, and I am the parent of Stephen Tsiorvas, who at the age of 10 lost his life along with two other sons of this community in a devastating accident that was like so many others preventable.

When I decided to come and speak with you today, I puzzled over what it was I most wanted to say. What of all the things I feel, believe and know were most important to relate to you. I wondered what words I could speak that would make the most impact, and inspire change so that this tragedy would never be repeated, so that other lives would be saved.

The first thing always on my mind is the depth of my sorrow and grief over Stephen's death, and how much pain his loss has brought my family. I could go on about that for hours. I could tell you how sometimes the sadness tears our hearts apart and drowns our spirits, but our grief is personal, as would be yours had you lost a child by any means. My sadness, my family's suffering can only serve to remind you how precious life is, how important it is that we love and protect our families and how easy it is to lose those we love. We need to take better care of our children, our loved ones and our neighbors. We need to commit ourselves again to making human safety a priority.

Most folks in Bellingham, many in Washington and some throughout the Nation realized again on June 10th, 1999, how unexpectedly dangerous and deadly our neighborhoods can be. The price of one human life is too great a price to pay for such a reminder. Safety measures can and must be taken. Taking a human life is not a business liability. It is murder. How many times over how many years must the greed and sloth of industry be allowed to play deadly games with human lives? Is that to be our future? Our children's future? Our children's children's future?

I need not remind the distinguished Committee that our government was formed of the people, by the people, for the people. Our democracy was based on that principle with a vision of the future very much on the minds of those drafting the rules for a balanced government. However, it does seem to me that we tend to live rather selfishly. With our individual lives and our individual pocket books and mind rather than the lives of our neighbors, and the lives of those yet to be born, this selfishness seems especially evident in those situations in which we allow the fat corporate wolf to manage the fat company wolves who then manage the hens in the hen house with the corporate agenda. While some wolves may be fine fellows indeed, a wolf is by nature a wolf. His agenda is based more on filling his stomach than on the welfare of the chickens.

Our environment, our habitat, our earth cannot continue to be raped, either by accident or by deliberate intent. If we are to sustain human life and animal life, if we erode our environment one small bit at a time, the human race will die, if not today, tomorrow; if not by fire, by ice; if not suddenly, then slowly.

The pipeline that burst here in this small city has been in the ground for a long time. Others will testify to that, I'm sure. The quality of that pipe and other pipes in this nation carrying volatile fuels is in question, but the impact of yet another disaster, the impact of yet another death is not.

Each accident wrecks havoc on the lives of residents, the lives of native animals, and the land itself. Each accident causes this nation, our neighbors, our families and our habitat irreparable damage. We know that toxic materials in our environment can poison us, leach into our food, pollute our drinking water and hover in our air, yet we continue to allow the wolves with the corporate agenda to manage not only the hens in the hen house but the hen house as well.

The size and growing complexity of our nation's needs does not abrogate our responsibilities as individuals, as a community or as a nation. Nor does a challenge demand that our complexity allow us to simplify our solution or grow lax in our diligence to monitor and enforce our regulations. Though the pipe is buried and out of our sight, the need for decisive action is abundantly clear. We can no longer hide from the fact that our behaviors and laissez-faire policies have direct detrimental effects on our future. By allowing inaction and by our own inaction, we become participants in the misdeeds of business. My baby died because of inaction. His death was preventable.

As a people, for the people, we must, must protect those lives in our charge. The lives of those yet unborn and the land which sustains humanity only for so long as humanity cares for it. Thank you.

[The prepared statement of Ms. Dalen follows:]

PREPARED STATEMENT OF KATHERINE DALEN

I wish to thank Senator Gorton, Senator Murray, and the distinguished members of the Committee on Commerce, Science, and Transportation for inviting me to speak today. My name is Katherine Dalen, and I am the parent of Stephen Tsiorvas who, at the age of ten, gave up his life, along with two other sons of this community, in a devastating accident that was, like so many others, preventable.

When I decided to come and speak with you today, I puzzled over what it is I most wanted to say, what of all the things I feel, believe, and know were most important to relate to you. I wondered what words I could speak that would make the most impact and inspire change so that this tragedy would never be repeated, so that other lives would be saved. The first thing always on my mind is the depth of my sorrow and grief over Stephen's death and how much pain his loss has brought my family. I could go on about that for hours. I could tell you how sometimes the sadness tears our hearts apart and drowns our spirits. But our grief is personal, as would be yours had you lost a child by any means. My sadness, my family's suffering can only serve to remind you how precious life is, how important it is that we love and protect our families, and how easy it is to lose those we love.

Additionally, I think I must speak to two issues. One, we need to take better care of our children, our loved ones, and our neighbors. We need to commit ourselves again to making human safety a priority. Most folks in Bellingham, many in Washington, and some throughout the nation realized again on June 10, 1999, how unexpectedly dangerous and deadly our neighborhoods can be. The price of one human life is too great a price to pay for such a reminder. How many times over how many years must the greed and sloth of industry be allowed to play deadly games with human lives? Is that to be our future as well? Our children's children's future?

I need not remind this distinguished committee that our government was formed of the people, by the people, for the people. Our democracy was based on that principle, with a vision of the future very much on the minds of those drafting the rules for a balanced government. However, it does seem to me that we tend to live rather selfishly, with our individual lives and our individual pocketbooks in mind rather than the lives of our neighbors and the lives of those yet to be born. This selfishness seems especially evident in those situations in which we allow the fat corporate wolf to manage the fat company wolves who then manage the hens in the hen house with a corporate agenda. While some wolves may be fine fellows indeed, a wolf is by nature a wolf: his agenda is based more on filling his stomach than on the welfare of the chickens.

Two, our environment, our habitat, our earth cannot continue to be raped, either by accident or by deliberate intent, if we are to sustain human and animal life. If we erode our environment one small bit at a time, the human race will die. If not today, tomorrow. If not by fire, by ice. If not suddenly, then slowly. The pipeline that burst here, in this small city, has been in the ground for a long time; others will testify to that I'm sure. The quality of that pipe, and other pipes in this nation carrying volatile fuels, is in question, but the impact of yet another disaster is not. Each accident wreaks havoc on the lives of residents, the lives of native animals, and the land itself. Each accident causes this nation, our neighbors, our families, and our habitat irreparable damage. We know that toxic materials in our environment can poison us, leach into our food, pollute our drinking water, and hover in our air. And yet we continue to allow the wolves with a corporate agenda to manage not only the hens in the hen house but the hen house as well.

The size and growing complexity of our nation's needs does not abrogate our responsibility as citizens or as a government. Nor does the challenge demanded by that complexity allow us to simplify our solutions or grow lax in our diligence to monitor and enforce our regulations. The need for stern decisive action is abundantly clear. We can no longer hide from the fact that our behaviors and laissez fair policies have direct, detrimental effects on our future, and all futures to come—on us, on our children, and on our children's children. As a people, for the people, we must, MUST, protect those lives in our charge now, the lives of those yet unborn, and the land which sustains us only for so long as we care for it.

It is time to act. We must not be deluded by false hopes or by denial; we are in danger; we must not allow the fact that we do not see the danger laying about in our front yard to make us unwary. If we do nothing we may lose our chance. And we must let our neighbors, throughout the nation, know of the danger that lurks three feet, two feet, one foot below the surface of our homes, our fields, our parks, and our schools.

Senator GORTON. Thank you very much, Ms. Dalen.

Mrs. Robinson?

STATEMENT OF MARLENE ROBINSON

Mrs. ROBINSON. I want to thank you for coming to Bellingham and thank you for inviting us here to speak.

On June 10th of last year, my 18-year old son, Liam, who had graduated from high school 5 days before, happened to be fly fishing in his favorite place, Whatcom Falls Park, a pristine piece of nature not 5 minutes from downtown. Just a week before, Liam had come home one evening from fishing the creek. While we ate dinner together, he told us about his excitement when he came upon a big otter swimming peacefully in one of the pools. He watched it for a long time. I'm sure that he looked for that otter on June 10th as he made his way down the creek. He was in a steep gorge with 230,000 gallons of gasoline spilled down the creek. The oxygen in the gorge was replaced by a 35-foot wall of hydrocarbon fumes. Liam was overcome within seconds. He fell into the foot-deep creek and drowned. A short time later, the gasoline and fumes exploded, sending the fire ball down the creek that killed Liam and Stephen and every other living thing in its path for a mile and a half.

We in Bellingham are now painfully aware of the danger that pipelines pose to every community in this nation. We have learned that what happened in Bellingham was not an isolated incident. The Federal Government has allowed the pipeline industry to be largely self-regulated. This has led to a pattern in the last 20 years of fuel transportation accidents. The pipeline industry will never have as its bottom line the health and safety of communities. It is up to communities, themselves, and therefore their public representatives and government agencies to insure that pipelines are safe.

The technology exists for pipelines to be safe. What we did not know before the pipeline ruptured in Bellingham but have learned at the price of our son's life is that what is lacking is regulation and enforcement. The Federal Office of Pipeline Safety has woefully, and over a long period of time, failed in its mandate. The Federal Government has not responded to years of unsafe pipeline practices and has at the same time prohibited local and state governments from protecting their citizens.

We in Bellingham are now working closely with many communities across the Nation who, like us, are educating themselves about the dangers posed to their citizens as a result of the lack of responsiveness of the Office of Pipeline Safety. We know that had OPS addressed this issue adequately in the past, our town would not be still reeling from loss. We are working together to make sure that no other community has to suffer a similar loss.

I no longer have children to protect. Nothing I do or say about this issue can bring Liam back. I do, however, consider it my privilege and my obligation to do what I can to protect the children of this and other communities. I need to impress upon you that it is not enough to make minor changes in pipeline safety regulation and to once again hand over the reins to OPS. Before June 10th, none of us in Bellingham had any idea that we needed to be experts in fuel transportation safety. We frankly didn't even know that we had a gas line pipeline running through the very heart of Bellingham under streets, past houses, schools and parks. We thought we had a Federal agency called the Office of Pipeline Safety, and we had faith that that agency was doing its job. We no longer have that faith.

I urge this committee to do what is necessary to protect the citizens of this nation from further avoidable and predictable tragedies caused by inadequate regulation, oversight and enforcement. My recent education has convinced me that we need two things. We need a Federal Office of Pipeline Safety that is staffed by committed expert servants who have the health and safety of communities as their bottom line. And we need a strong, well-funded citizens advisory council to insure that over time we do not return to business as usual.

Our children's deaths were not trivial. They were not an "acceptable risk." We easily have the capacity to protect our communities from just this so-called "accident." What I need from you, what every community in this country needs from you is action that will finally guarantee us an Office of Pipeline Safety that truly protects the safety of citizens across the Nation, and that will include citizens in local and state governments as effective partners in the national oversight of pipeline safety.

[The prepared statement of Mrs. Robinson follows:]

PREPARED STATEMENT OF MARLENE ROBINSON

On June 10th of last year, my 18-year-old son Liam, who had graduated from high school five days before, happened to be fly-fishing in his favorite place, Whatcom Falls Park; a pristine piece of nature not 5 minutes from downtown. Just a week before, Liam had come home one evening from fishing the creek. While we ate dinner together, he told us about his excitement when he'd come upon a big otter swimming peacefully in one of the pools. He watched it for a long time. I'm sure that he looked for that otter on June 10th as he made his way down the creek. He was in a steep gorge when the 230,000 gallons of gasoline spilled down the creek. The oxygen in the gorge was replaced by a 35-foot wall of hydrocarbon fumes. Liam was overcome within seconds. He fell into the foot-deep creek and drowned. A short time later, the gasoline and fumes exploded, sending the fireball down the creek that killed Wade and Steven and every other living thing in its path for a mile and a half.

We in Bellingham are now painfully aware of the danger that pipelines pose to every community in this nation. We have learned that what happened in Bellingham was not an isolated incident. The federal government has allowed the pipeline industry to be largely self-regulated. This has led to a pattern in the last twenty years of fuel transportation accidents. The pipeline industry will never have as its bottom line the health and safety of communities. It is up to communities themselves and therefore their public representatives and government agencies to ensure that pipelines are safe. The technology exists for pipelines to be safe. What we did not know before the pipeline ruptured in Bellingham, but have learned at the price of our son's life, is that what is lacking is regulation and enforcement. The federal Office of Pipeline Safety has woefully, and over a long period of time, failed in its mandate. The federal government has not responded to years of unsafe pipeline practices and has at the same time prohibited local and state governments from protecting their citizens.

We in Bellingham are now working closely with many communities across the nation who, like us, are educating themselves about the dangers posed to their citizens as a result of the lack of responsiveness of the federal government through the Office of Pipeline Safety. We know that had OPS addressed this issue adequately in the past, our town would not still be reeling from loss. We are working to make sure that no other community has to suffer a similar loss.

I no longer have children to protect. Nothing I do or say about this issue can bring Liam back. I do, however, consider it my privilege and obligation to do what I can to protect the children of this and other communities. I need to impress upon you that it is not enough to make minor changes in pipeline safety regulation and to once again hand over the reins to OPS. Before June 10th, none of us in Bellingham had any idea that we needed to be experts in fuel transportation safety. We frankly didn't even know that we had a gasoline pipeline running through the very heart of Bellingham, under streets, past houses, schools, and parks. We thought we had a federal agency called the Office of Pipeline Safety, and we had faith that that agency was doing its job.

We no longer have that faith. I urge this committee to do what is necessary to protect the citizens of this nation from further avoidable and predictable tragedies caused by inadequate regulation, oversight and enforcement. My recent education has convinced me that we need two things. We need a federal Office of Pipeline Safety that is staffed by committed, expert servants who have the health and safety of communities as their bottom line. And we need a strong, well-funded citizens advisory council to ensure that over time, we do not return to business as usual.

Our children's deaths were not trivial; they were not an "acceptable risk." We easily have the capacity to protect our communities from just this kind of "accident." What I need from you; what every community in this country needs from you, is action that will finally guarantee us an Office of Pipeline Safety that truly protects the safety of citizens across the nation, and that will include citizens and local and state governments as effective partners in the national oversight of pipeline safety.

STATEMENT OF BRUCE BRABEC

Mr. BRABEC. I am Bruce Brabec and I'm Liam's stepfather and I want to thank you for the opportunity to speak today and thank you for coming to Bellingham.

Marlene and I, as all the families, have received amazing support from this Bellingham community, and it's, you know, kind of maybe helped us to go on from day to day is all of the support we've had from friends and from people that we don't even know, and from people at large including the state.

A lot of times in conversation with people who are being supportive to us, they'll say to Marlene and I, I'm sure the others have heard it, too, people have said that they can't imagine what this must be like, what it must be like for us, and I thought, well, I can give you some sense of what it is like, especially for those in the room who are parents, and give you just a little bit of what it's like on a day-to-day basis for us.

First of all, just imagine that you're going to go home tonight and your child isn't home and never will be and then add to this each morning very early when the newspaper is delivered, it bumps on the front porch and you wake up because of that sound and you're reminded of when the police stepped on your porch and awakened you, also, and they had come to tell you that your son was found dead in Whatcom Creek and then add to this the experience that each time you go to a gas station to get gas in your car and when you catch the smell of the gasoline as you're filling it up, you imagine what it might have been like for your child as he was engulfed by a wall of gasoline vapor while fly fishing on Whatcom Creek, and then add to each time somebody tells you a story about your child, which we like to hear, but it also makes you think about what your child might be doing now if he or she were alive.

Well, now you might have a little bit more information about what it is like for us and those are just a few of the things that happen to us on a daily basis.

Last week Marlene and I decided to visit the site where Liam's body was found. We'd been there a few days after his death and we weren't sure we could find it again and at that time we'd seen how badly damaged the creek was and we wanted to go back and some friends in the police department who knew where the site was offered to take us back and we wanted to go back because we wanted to have a good bearing on the site in the future, so we could visit it on our own.

I planned to stand there and imagine Liam fishing in this one beautiful canyon, and I imagined that in the future not even at that point it would be a site where I would be able to reflect joyfully about Liam, because I knew how much he loved fishing in that canyon. How often I'd come home, and there'd be a note that he had gone fishing, and I knew which creek he was fishing on.

Well, the experience of going to the creek last week was quite the opposite. I saw a burned out canyon, which as much as I expected it, it was very difficult to see again, the burned trees, the bare banks, the cracked rocks from the heat of the explosion, and the downed trees in the water, and worst of all, I saw Liam floating face down in the creek and I saw the part of his body that was out of the creek that was charred by the blast, because I know that that's how they found him.

Now, I'm not bringing this up to upset people, but really I bring it up to make a plea that Liam's death not be in vain and to plead that other families not have to be condemned as we are to this kind of experience and to these memories. The loss of our son—because of the loss of our son we are certainly biased about the importance of stricter regulation and accountability, but we believe that everybody should be similarly biased by our experience.

I would like to bring another voice into the room and that's a voice that's missing today and that's Liam's, Liam's voice. If I were Liam, and I hadn't been fishing that day, and I was alive, I would hope that I was invited to speak today, because I would tell you how often I had walked that creek from the mouth all the way up to its source. If I were Liam I would tell you how many of my thoughts had been scoured from that canyon by the blast, thoughts I shared with the creek as I fished, thoughts about school, thoughts about my parents, about girlfriends, about my place in the universe. If I were Liam I'd tell you how many fish I'd caught and released, and I would recount with pretty good sound effects and visualizations my attempts and success at landing fish, and if I were Liam, I would be telling you today that much more needs to be done regarding pipeline regulation, to not only prevent damage, injury or death, but also to support the living, to protect our places of refuge, to protect our neighborhoods, to protect our homes, to protect our families. If I were Liam, I would most strongly ask that you tighten the accountability of the Office of Pipeline Safety, that you support it with the funds to do its job, but that you consider cleaning house to get staff who are dedicated to be watch dogs of public safety, not just supporters of pipelines and oil companies. If I were Liam, I would most strongly urge you to allow states and especially Washington State to regulate pipelines in addition to Federal regulation, and if I were Liam, I would urge you to support a well-funded citizen advisory group to provide regional oversight to improve safety and to prevent oil spills, and if I were Liam, and I would be not standing up and moving over you at my over six feet height, and I would be volunteering to be the first one on that advisory committee for this region, because I would know how important it is, and I'd want to do what I could to be helpful.

Marlene and I again want to thank you for the opportunity to present our thoughts, our feelings today. We appreciate your serious intentions. Let's join ours and that is to take steps to prevent similar accidents from occurring in the future.

[The prepared statement of Mr. Brabec follows:]

PREPARED STATEMENT OF BRUCE BRABEC

I am Bruce Brabec. I am Liam's step father.

Marlene and I have received amazing support from the Bellingham community. And in conversations many people tell us that they can't imagine what it must be like for us.

Well, to give you some sense of what it is like, especially for the parents in the room . . .

- Imagine going home tonight and your child isn't home . . . and never will be.
- Add to this—each morning, very early, when the newspaper being delivered thumps on the front porch, you wake up and are reminded of when you were awakened by the police stepping on your porch—who came to tell you that your son was found dead on Whatcom Creek.
- Then—add the experience of each time you go to a gas station to get gas for your vehicle, and you catch a smell of gasoline—you imagine what it might have been like for your child as he was engulfed by a wall of gasoline vapor while flyfishing on Whatcom Creek.
- Then add that each time someone tells you a story about their child, you think about what your child might be doing now—if he were alive.

Now you might have a bit more information about what it is like for us.

Last week, Marlene and I decided to visit the site where Liam's body was found. We had been there once, a few days after his death, and weren't sure we could find it again. We had seen then how badly damaged the creek was. This time, we were accompanied by some friends in the police department who knew the site. We thought it would be good to get a better bearing on the site so that we could visit it later on our own. I planned to stand there and imagine Liam fishing in this once beautiful canyon—I imagined it as a site where I would be able to reflect joyfully about Liam as I knew how much he loved fishing in that canyon . . . Well, the experience was quite the opposite. I saw the burned out canyon, the burned trees, the bare banks, the rocks cracked by the heat generated during the explosion, the downed trees in the water. And I saw Liam floating face down in the creek and the part of his body out of the water charred by the blast.

I am not bringing all this up to upset people, but to make a plea that Liam's death not be in vain, to plead that other families not have to be condemned as we are to these kind of experiences and memories. Because of the loss of our son, we are certainly biased about the importance of stricter regulation and accountability, but we believe everyone should be similarly biased by our experience.

I would like to bring another voice to the room a voice that is missing . . . Liam's voice.

If I were Liam . . . and I hadn't been fishing that day . . . and I was alive . . . I would hope I was invited to speak today. I would tell you of how often I walked that creek from the mouth all the way up to its source. If I were Liam, I would tell you how many of my thoughts had been scoured from that canyon by the blast—thoughts I shared with the creek as I fished—thoughts about school, about my parents, about girlfriends . . . thoughts about my place in the universe. If I were Liam, I would tell you about the many fish I caught, and released, in that canyon. And I could recount with good sound effects and visualizations my attempts and success at landing a fish. If I were Liam, I would be telling you today that much more needs to be done regarding pipeline regulation to not only prevent damage, or injury, or death, but to also support the living, to protect our places of refuge, to protect our neighborhoods, to protect our homes.

If I were Liam, I would most strongly ask that you tighten the accountability of the Office of Pipeline Safety—that you support it with the funds to do its job, but that you consider cleaning house to get staff who are dedicated to being watchdogs of public safety, not just supporters of pipeline and oil companies.

If I were Liam, I would most strongly urge you to allow states and especially Washington State to regulate pipelines in addition to the federal regulations.

If I were Liam, I would urge you to support a well funded citizen advisory group to provide regional oversight to improve safety and prevent oil spills.

And, if I were Liam, I would volunteer to be on that first advisory committee for this region.

Thank you for the opportunity to present our thoughts today. We appreciate your serious intentions which join with ours—to take steps to prevent similar accidents from occurring in the future.

Senator GORTON. Well, we thank each of you who have gone through great difficulty and who are here today, you know, under a great deal of stress for the kind of insight into these challenges that only each of us can know.

I've only one or two brief questions, and I think Mrs. Robinson has already mostly answered the question.

How many of you before this accident even knew there was a pipeline through the park with these hazardous materials going through it? You said you were not, Mrs. Robinson?

Mrs. ROBINSON. No.

Senator GORTON. Mr. King, did you know it was there?

Mr. KING. I knew the pipeline was there. I really, I never paid any attention as to what went through it, because I felt that, you know, they were taking care of it.

Senator GORTON. What do you think——

Mr. WILLIAMS. Senator?

Senator GORTON. I'm sorry. Do you want to answer the question?

Mr. WILLIAMS. Yeah, I lived in the community for 10 years.

Senator GORTON. Uh-huh.

Mr. WILLIAMS. And I knew the pipeline was there, also, but I couldn't find it, and when I would ask neighbors what was in it, nobody knew.

Ms. DALEN. No.

Mr. WILLIAMS. Nobody knew what was being transported.

Senator GORTON. Did you know precisely where it was?

Ms. Dalen You can see it.

Mr. WILLIAMS. You can see where the right-of-way is in some of the areas, but once again, no one knew what was being transported in that pipeline.

Ms. DALEN. And it is literally out of the ground and goes across that black pipe. It goes across from one side of the creek in the little gully that the creek has made to the other. You can see it. So you assume it's water or something.

Mr. KING. My oldest son, Jason, fished like Liam Wood every inch of that creek for 15 years. We've lived there for 22 years. He's probably stood on the pipe and fished off the pipe as a little boy. He said it was—there was a greater chance of him being killed back in there than Wade.

Senator GORTON. You've already, I think, each of you in your testimony told us what you think we ought to do, but one question in that connection, what's the best way to see to it that people in the future here in the State of Washington or anywhere else who live in the vicinity of these pipelines are made aware of the dangers they pose and can participate before something happens in seeing to it that they're safe? Do any of you have a thought on that subject?

Part of Senator Murray's bill, of course, is public information and public knowledge. Would that be important to you?

Mr. KING. I think that the most important thing that needs to come out of this is the pipeline industry has no intention of doing any preventative maintenance on the pipelines at all. Are we going to eliminate spills? Probably not, but if we have a zero tolerance, and we did get to zero spills, wow, wouldn't that be something? The world isn't perfect, but does that mean that we don't strive for perfection?

Mrs. KING. It wouldn't have mattered that day if, what we knew. After it had happened and the oil had spilled, it wouldn't have mattered who knew, who we called, what we did. It was going to happen. This shouldn't happen. I sit here and I listen to all of this and I get angrier and angrier and angrier. If this is properly maintained, regulated, whatever you want to call it, this won't happen again. We won't have to have numbers to call. We won't have to be looking out. This isn't up to the citizens. This is our government that should be doing this, and I don't care about an 800 number to call if I detect a gas leak. If this is properly handled in the future, it can't be anything but better. Anything is going to be better than what's gone on in the past, which is nothing.

Mr. KING. Hyman Rickover who is an admiral in the Navy had a zero tolerance for nuclear accidents. There aren't very many accidents in the navy concerning nuclear power, because he had a zero tolerance.

Ms. DALEN. I also the wanted to say something. I believe that there are several avenues that we can take to inform the public. One of the things that I've been very concerned about is the lack of information. Well, there has been some information, but through the news media making sure that the country understands that these things are not just out in the farmland which is bad enough, not out, just out in the ranges which is bad enough, but within feet of public schools, in the middle of parks, down the back side of people's yards. They need to know where it's at. So I ask the media and the government to demand that the pipeline companies let us know where their pipes go, No. 1.

I recognize near the beginning of this tragedy it was very difficult for us to get Olympic to be forthcoming with their maps. It seemed that they didn't exactly know where their own pipes were going. Well, maybe that might be a little off. Maybe they were trying to put us off, but nevertheless, they need to be forthcoming. We need to have that information available, and it can be printed. People all over the country in every type of newspaper from the New York Times down to the, you know, Pullman Herald can find out in their newspapers using the Internet and making these things available for people who want to take an active role, be actively involved in it, making sure that every owner, every property owner knows that there's, if there's a pipeline going through or by their property, and every time those homes are bought or sold, every apartment bought or sold, every parking lot bought or sold, every business bought or sold that goes near a pipeline, make that absolutely a part of the deal. They need to know.

Speaking of the nuclear industry, I happen to be under the impression that they have pipes, and they have smart pigs that are a heck of a lot smarter than the pigs they're using to chase down our pipe leaks in our pipes, and I think we ought to start using them. I don't care if they cost a little bit more. Excuse me. I'm paying enough for gas right now. I'm willing to pay an extra few bucks to save my neighbor's children. I'm willing. I'm willing, and I think we ought to start making use of multiple industrial safety measures and asking other people how they keep their pipes safe.

Senator GORTON. Thank you.

Senator Murray?

Senator MURRAY. Mr. Chairman, I don't have any additional questions. I just want to thank this panel for being so courageous and for coming and sharing your stories with us. I know how difficult it is. I know how much each of you have helped to educate all of us, and I hope we can take your courage back to the other Washington to convince senators and legislators from across the country that what happened in Bellingham last year could happen in their community today unless we toughen many of these laws and go through with many of the recommendations you've talked about.

So just personally, thank you very much for coming here and sharing your stories.

Mr. Chairman: Thank you all.

Mr. KING. May I make one——

Senator GORTON. Sure, of course you can.

Mr. KING. I asked this back in Congress, why is Olympic Pipe Line still operating south of Bayview station with their eight employees still at the switch pleading the Fifth Amendment? Why?

Senator GORTON. That we hope we learn from people who are going to testify here later on in the day.

Mr. KING. Well, the problem is there is no logical reason why they're still operating, and everybody I ask that question, they get stumped. Nobody wants to, nobody wants to address it.

Can you go back to Congress and ask among your other senators why are these people still operating south of Bayview station? All we're asking for is to make the pipeline industry and particularly Olympic Pipe Line accountable, accountable for what they did.

Senator MURRAY. Thank you.

Senator GORTON. Thank you all very much.

Governor Locke is here. We've given him the difficult assignment to testify next, but Governor, we're honored to hear from you.

Governor Locke.

STATEMENT OF HON. GARY LOCKE, GOVERNOR, STATE OF WASHINGTON

Governor LOCKE. Thank you very much, Senator Gorton and Senator Murray. Thank you, Senator, for holding this field hearing to allow the people of the State of Washington to express their concerns about the safety of pipelines that as parents so courageously testified just a few minutes ago, are in our parks, are in our yards, are next to our schools, in the midst of our own communities.

On June 10th, 1999, Liam Wood, recent high school graduate, Wade King, Stephen Tsiorvas, they died in a tragedy that never should have happened. There can be no more children dying, not in Washington State, not anywhere in America. We owe it to them to make sure that a similar tragedy never occurs again throughout the United States of America.

Since that horrible tragedy on June 10th, we've all become so aware of the risk surrounding these vital pathways so central to the distribution of fuel. We realize we need pipelines to carry the natural gas and the petroleum products that fuel and power our cars and our trucks, that warm our homes and keep our factories humming, but for virtually all of us, the risk posed by the pipelines were out of sight, out of mind.

Immediately after this tragic event, I formed a task force to evaluate our pipeline regulatory system and our response capability, and that team worked incredibly hard for 5 months, and Mayor Asmundson of Bellingham was part of that team, and they spent their time analyzing the existing regulatory system and response capability and preparing recommendations.

One of the first things they and the people of the State of Washington learned was that our state has virtually no control, no control over the pipeline that leaked the gasoline that exploded. This interstate pipeline as well as six other interstate petroleum and natural gas pipelines is solely under the authority of the Federal Office of Pipeline Safety, and until Senator Murray succeeded in obtaining an Office of Pipeline Safety position for our state, the entire western region of the United States had only 13 inspectors, three located in Alaska, 10 others for the entire western United States. That is simply inadequate and is a recipe for disaster.

When the task force finished its work, they issued a report that included 30 strong and solid recommendations for improving pipeline safety, not just in our state but all across America. I have endorsed these recommendations and working with our state legislature and other Governors and the members of our congressional delegation, we're trying to turn them into reality, but I want to thank State Senator Harriet Spanel to my right, and State Representative Kelli Linville for prime sponsoring the legislation in Olympia that just a few days ago our legislature passed: the Washington State Pipeline Safety Act which will truly strengthen the monitoring and the prevention of accidents in our state.

The bill will also initiate studies by the State Fire Marshal of the training and the equipment needs facing communities that lie along these pipelines, but we need more than anything else to have the Federal Office of Pipeline Safety set tougher standards and to have more stringent inspections of all pipelines, natural gas and volatile fuels. We simply must have the Federal Pipeline Safety Act amended to allow states to adopt and enforce standards stricter than Federal standards when doing so would not interfere with interstate commerce.

The Office of Pipeline Safety is woefully understaffed and has not kept abreast with the latest developments to ensure pipeline safety. Therefore, the Congress must step in and insist on a tougher regulatory stance by the Federal agency, and in fact, allow the states to go even farther to protect our own citizens from such tragedies. Even before that occurs, the Office of Pipeline Safety must grant authority to the states for interstate pipeline oversight using current or future Federal standards as it has done with four other states. Four states now have been delegated the authority by the Federal Office of Pipeline Safety to be in essence an arm of the Federal Government given the fact the Federal Government has been so lax. Washington has been asking for this delegation of authority as have many other states including Virginia and Arizona.

We want the ability to help the Federal inspectors do their job, but it should not take deaths before the Federal Government says yes.

Just today we received a letter from the Federal Office of Pipeline Safety indicating that it is willing to delegate authority to the State of Washington.

Senator GORTON. Do you have a copy of that letter——

Governor LOCKE. Yes.

Senator GORTON. —that we can put in the record?

Governor LOCKE. Yes, I do—but they're saying yes to our request, because we've had three deaths. What about all the other states that are looking for similar authority where there have been massive spills and leaks, but without a death? The Office of Pipeline Safety should not have to wait for more deaths in other parts of American before granting similar authority.

I have sponsored and the National Governors Association has adopted a resolution supporting stronger Federal pipeline safety as well as increased state involvement. We've been working with Senator Murray and Representative Metcalf and Representative Inslee to promote their bills that will No. 1, push the Office of Pipeline Safety to adopt stronger standards it should have enacted years ago, and that will two, authorize states to go beyond the Federal standards on issues like training, certification, leak monitoring, and accident preparedness, and I want to thank you, Senator Gorton, for cosponsoring Senator Murray's bill. We appreciate the time that you're spending to give the people of the State of Washington the opportunity to indicate their concerns about these tens of thousands of miles of pipeline running through their communities. We've been pushing hard to get the Federal Government to give us that authority. They're now beginning to respond, but we're going to have to continue to push, today, tomorrow, every day.

We need to make sure that the legacy of Liam Wood, Wade King, and Stephen Tsiorvas will be one of protection for all the people all across America. Thank you very much.

[The prepared statement of Governor Locke and letter referred to follow:]

PREPARED STATEMENT OF HON. GARY LOCKE, GOVERNOR, STATE OF WASHINGTON

I want to express my appreciation to Senator Gorton and Senator Murray for holding this field hearing to allow the people of Washington to express our concerns about the safety of the pipelines that lie below our towns and cities.

Since the terrible tragedy on June 10, 1999, we have become all too aware of the risks surrounding these vital pathways so central to our fuel distribution network. We realize we need pipelines to carry the natural gas and petroleum products that power our cars and trucks, warm our homes, and keep factories humming and airplanes flying. But we also know that for virtually all of us, the risks posed by the pipelines were "out of sight, out of mind" before the explosion in Whatcom Falls Park.

Immediately after this tragic event, I formed a task force to research our pipeline regulatory system and our response capability. That team worked incredibly hard for 5 months, analyzing the existing regulatory system and response capacity and preparing recommendations for improvement. They consulted with experts from the federal government—including the National Transportation Safety Board and the Office of Pipeline Safety—from other states, including Minnesota and California that operate their own interstate pipeline safety oversight programs—and from a variety of interest groups.

One of the first things they—and the people of Washington—learned was that our state exercises almost no control over the pipeline that leaked the gasoline that exploded. This interstate pipeline—as well as 6 other interstate petroleum and natural gas lines—is solely under the authority of the federal Office of Pipeline Safety. And until Senator Murray succeeded in obtaining an OPS position for our state, the entire Western Region had only 13 inspectors.

When the task force's work was done, they issued a report that included over 30 strong and solid recommendations for improving pipeline safety. I endorsed those recommendations and have been working with our state legislature, with other Governors, and with members of Washington's congressional delegation to turn them into reality. I would like to acknowledge the hard work done by the legislature on this subject, especially Representative Linville and Senator Spanel.

Since December:

- The Legislature adopted the Washington State Pipeline Safety Act, which will strengthen our "Call-Before-You-Dig" program to reduce third-party damage to pipelines and will promote comprehensive mapping of pipeline locations to inform local governments who must make building permit decisions and prepare for pipeline accident response. The bill will also initiate studies by the State Fire Marshal of the training and equipment needs facing communities that lie along pipelines.
- The Legislature also adopted a Joint Memorial to Congress and the President that cited the work of the task force and urged you to amend the federal pipeline safety act to allow states to adopt and enforce standards stricter than federal standards when doing so would not interfere with interstate commerce. Even before that occurs, the Memorial asks the President to direct the Office of Pipeline Safety to use existing law to grant authority to states for interstate pipeline oversight.
- I sponsored, and the National Governors Association adopted, a resolution promoting stronger pipeline safety and committing the NGA to work with Congress on legislation to achieve that objective. This parallels a similar resolution that several Washington cities successfully promoted to the National League of Cities.
- I have been working with Senator Murray and Representative Metcalf to promote their bills that will push the Office of Pipeline Safety to adopt the stronger standards it should have enacted years ago and that will authorize willing states to go beyond federal standards on issues like operator training and certification, leak detection, and accident preparedness.
- Senator Gorton, I want to thank you for co-sponsoring Senator Murray's bill. I appreciate your recognition that pipeline safety is of paramount importance to the people in our state and I want to do everything in my power to work with you to ensure safe lines.

I am confident that with the concerted effort of my Administration and our entire congressional delegation we can change the federal law and institute a substantially stronger program here in Washington. That will enable us to give our citizens the peace of mind they deserve that the pipelines beneath our state are operated and maintained to protect public safety and our environment. This will be the legacy of Liam Wood, Wade King and Steven Tsiorvas.

Thank you very much.

U.S. DEPARTMENT OF TRANSPORTATION, RESEARCH AND SPECIAL PROGRAMS ADMINISTRATION
Washington, DC.

Hon. GARY LOCKE,
Governor of Washington,
Olympia, WA.

Dear Governor Locke:

I am writing to transmit to you two draft Interstate Pipeline Transportation Agreements: one for hazardous liquids and one for gas.

Since our meeting in Washington, DC, in late February, my staff and I have been working to find a way to authorize the state to serve as our agent in conducting interstate pipeline oversight. Based on our assessment, we have developed these draft agreements.

I must convey to you our perspective on implementation of these agreements, should you choose to accept them. The Office of Pipeline Safety has significant concerns about having the State of Washington's pipeline safety program housed in more than one agency. This concern arises for two main reasons:

- With a single pipeline safety agency, the critical mass of expertise can be established to contribute to activities effecting both liquid and gas pipelines, for example, metallurgical sciences, computer operations management, mapping, etc.

- While I understand that Washington is prepared to provide adequate funding, it is inevitable that additional resources will be required to duplicate expertise in two locations. As I remarked when we met, I am concerned about the state's ability to meet the resource needs associated with interstate agent status and do not want that problem exacerbated by unnecessary inefficiency.

As you know, we have had an effective working relationship with the Utilities and Transportation Commission. Over the years we have worked with them on their certified intrastate liquid and gas programs. This agency has consistently achieved the highest level of funding we could make available as a reflection of their successful performance of their responsibilities. We would be concerned about any lapse in this performance during a transition to a second agency.

Having reviewed the bill adopted by your Legislature, I must note that we would have some concern if there were any legal uncertainty surrounding the agency responsible for interstate liquid pipeline oversight. I would hope that a means can be found to avoid this problem.

I thank you for your strong interest in promoting pipeline safety and look forward to working with you to achieve that objective.

Sincerely,

KELLEY S. COYNER,
ENCLOSURES

Senator GORTON. Thank you, Governor.

You have two distinguished members of the legislature with you, and in light of the letter that you've just received from OPS, is the legislation that they've sponsored and that I take it that you're about to sign, is that adequate to meet what you consider the responsibilities of the state to be?

Governor LOCKE. The legislation that was sponsored by Senator Spanel and Representative Linville does enact most of the recommendations of the task force that I formed several months ago. The Office of Pipeline Safety, however, has expressed a few concerns that are more technical issues dealing with whether or not the duties that the state would assume with the delegation of authority that's being proposed by the Federal Government might be diluted if it were in several agencies, and so a letter from Ms. Kelly Coyner is strongly suggesting that the oversight be continued to be housed in the current Utilities and Transportation Commission. The Office of Pipeline Safety is concerned that in our legislation, it's being bifurcated into two different agencies, but I think that this can be solved. So I see no reason why the delegation of authority can't proceed, so that the state can just help enforce the Federal standards, using the same Federal standards or whatever the Federal standards might be, and hopefully those Federal standards will be toughened up, but we're ready to do it. We're willing to use our own state resources to do it and to bring our people into the task, and so I think with these two legislators here, if there are any further technical amendments or corrections that need to be made we can address those in a special session now under way in Olympia.

Senator GORTON. You've anticipated my next question. While Senator Murray's bill that has my support calls for Federal assistance to the states, that's not a certainty by any stretch of the imagination. You feel that you can secure adequate resources to deal with this delegation here?

Governor LOCKE. Yes, we believe that we'll be able to find the dollars because this is of such utmost importance for safety in communities throughout the State of Washington, we will make it work.

Senator GORTON. Do you think that Senator Murray's bill delegates the state and local governments an appropriate amount of authority?

Governor LOCKE. Yes, I do, and I very much support Senator Murray's bill and also Representative Metcalf's bill. No. 1, it would require the Federal Government to set even tougher standards, and No. 2, it enables or clearly indicates to the Federal agencies that they must enlist the support of the states, and those states that are willing to do it, like the State of Washington should not be thwarted in their effort. We should not have to wait until there's a death before the Office of Pipeline Safety is willing to consider that delegation and sharing of responsibilities with the state.

The Federal Government, excuse me, the Office of Pipeline Safety, has been woefully negligent in this area. There are many states ready and willing to step up to the effort to help out using our own nickel, our own resources and the Federal agency should not be reluctant to engage in that partnership with the states.

Senator GORTON. We have at least heard the rumor that the Administration may want to take back that right to delegate power to the states even in the limited amounts that it already has. I know neither of us is going to approve of anything like that, and you feel that the states have a potentially a far greater scope, not a lesser scope than the present statute allows?

Governor LOCKE. I firmly believe that it's thoroughly proper for the Federal Government to set minimum standards, and that the states should have the latitude to set tougher regulations on top of that to really protect our citizens just like in the area of tanker safety through our straits and so forth. The Federal Government should help set a minimum floor, but the states should not be precluded from enacting tougher regulations to deal with any particular circumstances in their communities, and clearly, given our environment and the fact that these pipelines are going right through the heart of communities, we should be able to ensure our citizens that we're doing everything we can, especially since the Federal Government or the Federal agency has been woefully inadequate in this area.

Senator GORTON. Thank you.

Senator Murray?

Senator MURRAY. First of all, Governor and Harriet Spanel and Kelli Linville, thank you for the tremendous amount of work you did in working this through this legislative session, and I am delighted to hear that you've received a letter from Kelly Coyner on a state delegation of authority. I'm glad that they have responded to you.

I have a couple of questions. They have agreed to delegate the authority. It sounds to me like you need to deal with the concern of the split delegations or split regulations. You will be able to deal with that in special session so that that concern can be addressed?

Governor LOCKE. I believe we can address that, whether administratively or through legislation in Olympia, if necessary. What

Ms. Coyner has indicated is they're sending us some draft agreements and our lawyers are working on it, and our administration people are working on it. This comes about after I came back to Washington, D.C., about 2 weeks ago and I had a very long meeting with Ms. Coyner and Secretary of Transportation, Rodney Slater, and impressed upon them that in seeking this delegation, we were not seeking Federal funds, that we were not using this as a pretense for greater Federal funds, that we're prepared to do this on our own nickel, using our own resources, our own people, because this is so incredibly important.

Senator MURRAY. I want to make it clear, what they're able to delegate to you is to regulate current Federal standards, and unless we raise the Federal standards to higher training and certification standards and inspection standards, you don't have much more authority than is currently out there, and we need to do that as part of the next step in this process.

Governor LOCKE. That is correct. Again, this delegation would be basically letting the states be almost an extension of the Federal agency, but we all know that the Federal standards are too lax. They need to be tightened up, and so we welcome your legislation to insure that the Office of Pipeline Safety is enacting and promulgating the most stringent requirements with respect to integrity of the pipes, inspection, enforcement, and monitoring, but then states should also have the latitude to go even farther than the Federal Office of Pipeline Safety rules and regulations.

Senator MURRAY. Which is what we do in our legislation.

Governor LOCKE. That's right.

Senator MURRAY. You indicated or responded to the question about resources which is absolutely a critical one. This is obviously a very hotly discussed topic here in the State of Washington right now. Five years down the road from now, it may not be if we all do our jobs correctly, and we all hope we do. How do we know that 5 years from now, 10 years from now when leadership changes and other people are in place that they will continue to commit those same resources and follow the same standards that we are all so adamant about today?

Governor LOCKE. Well, I can't guarantee the caliber of the people at the Federal agency or the number of inspectors throughout the states, throughout the western United States, and that's why I think it's important that each state be granted the authority if they so choose to be an extension of the Federal Government, and that's why it's important that states also have the latitude to enact tougher standards knowing that their citizens, their children and their civic leaders will be demanding greater scrutiny. Basically, you know, we're closer to the people here, and if we at the state level are not doing the proper job, those citizens will have greater success in making sure that the job is done as opposed to trying to lobby or convince the Federal Office of Pipeline Safety.

Senator MURRAY. My concern is just that when budgets get tight and issues become more difficult, will resources remain there for the state to be able to oversee that and that will be the responsibility of all legislatures at the time, I assume, and Harriet, you wanted to add something?

Senator SPANEL. In the bill we have the ability to look at funding in the form of fees from the users.

Senator MURRAY. Kelli, did you have any comments?

Representative LINVILLE. We also had anticipated a potential problem with the division of the two departments and hopefully had signalled our willingness to revise the legislation for the fact that caused a problem with the delegation, and I did bring a copy of our legislation and a copy of the summary and was very happy to hear both of your comments. I think we included a lot of your interests in how you would change the Federal law in our state law, and be very grateful and speaking on the fact of, on behalf of the people in the community if in fact these recommendations that we've worked so hard to put together would get included in some Federal legislation that ups the standards, because as you said if we get the delegation, the delegation we get now is only to inspect to the Federal standards. I believe we can still do a better job in Washington State with the Federal standards than the Office of Pipeline Safety has been doing, but I would implore you to develop greater standards. I think the citizens of this community are demanding prevention not response, and stricter standards are going to be what takes care of that.

Senator SPANEL. I would just add—in listening to testimony on the bills in Olympia, it's more than just Bellingham at this point. It is all up and down the I–5 corridor, and I think this is a bill that shows that a lot of people really supported it. It passed unanimously out of the senate demanding exactly the things that Governor Locke has stated: that we want tougher rules on the Federal level, want to be able to make tougher rules on the state level, and we do want the authority to enforce them. I understand your question on funding, and that is why we do know there has to be another source, but I, for one, would never turn down Federal dollars in the future.

Senator MURRAY. Well, thank you very much to all three of you for the tremendous amount of work you've done. You've done a good job. We now have a responsibility to do ours.

Thank you very much.

Senator GORTON. Governor, thank you. I do appreciate, you've now become a national spokesperson and a national leader on this subject, and your words are being heard in Washington, D.C., and Senator Spanel, Representative Linville, congratulations. You work faster than we do.

Senator SPANEL. We have shorter sessions.

Governor LOCKE. Thank you.

Senator GORTON. The next panel is a group of local officials along that I–5 corridor, Mayor Asmundson, Mayor Tanner, Deputy Mayor Marshall and City Manager, Joe Hoggard.

Mayor Asmundson, I share the compliments that Senator Murray began this hearing with and would also like to thank you for making this facility available for a hearing. We really filled up city hall.

Mayor ASMUNDSON. Yes, indeed, you have filled up city hall.

Senator GORTON. And we are, since you're our host, you will start this panel group testimony.

STATEMENT OF MARK ASMUNDSON, MAYOR, BELLINGHAM, WASHINGTON

Mayor ASMUNDSON. Thank you very much, Senator Gorton, and also to Chairman McCain and to other members of the Commerce Committee conducting the field hearing today. I appreciate the opportunity to testify before the Committee, and I'm very grateful to see Senator Murray with us, also, so thank you.

The City of Bellingham, its citizens and I think the other mayors and representatives of citizens up and down the I–5 corridor appreciate the opportunity to discuss pipeline safety. I'm going to try and not repeat much of what you've heard today, and much of what I've spoken with you both about in the past, but I will try to focus on some issues that I think just do not leave me in the course of my learning about pipelines and interstate pipelines in this country.

The Olympic pipeline which travels through Bellingham goes through 21 cities in Washington, cities with a combined population of nearly 800,000. It also goes through very many environmentally sensitive areas, lakes, aquifers, salmon bearing streams and critical wetlands habitats.

Following our June 10th disaster, I began a period of intense involvement in learning to understand the fuel pipeline industry, its regulatory regime, and the circumstances that might have led up to the horrible event that occurred here in Bellingham, and I discovered that this event was not unique. As Senator Murray said earlier, this is not a unique occurrence. Based on the history of the pipeline industry in the United States, I concluded that this event could have been expected. In fact, given the current state of affairs involving interstate fuel pipelines in America, the issue is not will pipelines leak, the question is when will they leak, where will they leak, and how bad will the harm be.

Over the course of the last two decades, there's been a long history of disaster after disaster. Sometimes there's serious injury. Sometimes there's death. There's always serious environmental damage. While these continuing occurrences cause alarm, what causes even more alarm, I think, is the answer to the question why.

The Federal Government has preempted the regulation of pipeline safety. The Office of Pipeline Safety in the Department of Transportation has the responsibility of carrying out the Federal Government's protection of the public from the hazards of interstate fuel pipelines, but the Office of Pipeline Safety has done a remarkably poor job over the years, remarkably poor job. In part, this is because of limited resources. However, my observations of the record indicate to me that the fundamental problem with the Office of Pipeline Safety is a lack of will on the part of the Office of Pipeline Safety. The OPS does not aggressively and effectively pursue public safety. It seems intent on making certain that it takes no action that will be unacceptable to the industry they're charged with regulating. So much so that the Office of Pipeline Safety fails to comply with the explicit mandates of Congress as you pointed out earlier.

The rules under which the pipeline industry operates simply must change. Clearly the Federal Government must establish minimum levels of safety that must be followed by this industry throughout the country. However, the Federal Government's legitimate concern and need to protect interstate commerce does not preclude states having a meaningful role. A partnership can and should be established between the Federal Government and states whereby states can protect their citizens without interfering in interstate commerce.

Interestingly, if the Office of Pipeline Safety were doing its job, we wouldn't have a debate, because we wouldn't be asking for delegation of authority to the states to protect our citizens, because it would have been done, but it has clearly not been done. It's not even come close to having been done, and it is for this reason that I personally and with the enthusiastic support of the community of Bellingham support Senate Bill 2004, prepared by Senator Murray and cosponsored by you, Senator Gorton.

I believe that this bill together with similar efforts that the house of representatives set forth in House Bill 3558 by Representative Metcalf will accomplish changes that are necessary to ensure that not only will we have a strong, viable fuel distribution system in America, but we can have a strong, viable fuel distribution in America that does not endanger our environment, does not kill our children and allows our citizens living near pipelines to go to sleep at night without anxiety about their personal safety.

Now, I'd like to talk a little bit about pipeline safety from the perspective of what it means in practical human terms as a lifetime resident and citizen of Bellingham, and as a person who happens to be the mayor right now. Fundamentally pipeline safety is not about legislation. It's not about Federal bureaucracy. It's not about concepts. It's about people, and it's about the environment.

When a pipeline rupture occurs, it doesn't happen in a committee hearing. It doesn't happen in an agency office. It happens in a neighborhood or in a park or a wetland, farm or over a stream. The effects of the rupture are not theoretical. They're not abstract. They're very real, and in Bellingham's case it meant attending the funerals of three wonderful boys in 1 week along with a community of thousands of mourners. It meant the disruption of our water supply for a quarter of our citizens while alternate facilities were being developed. It meant the indefinite delay in the restoration of a salmon habitat restoration project. It meant children were afraid that their neighborhoods might blow up. It meant anguish, questioning, grieving and in most cases a totally unsatisfied search for rational answers to the question why.

In Bellingham's case it means great frustration on the part of locally elected officials like myself when asked by citizens, "What will you do to make sure that we are safe?" The answer, "I will do the best I can, but my hands are tied by Federal law," is not very satisfactory.

The disaster in Bellingham was not unique. As Senator Murray pointed out, they occur all over the country regularly. In one way though, Bellingham was unique, because Olympic Pipe Line Company had failed to keep its franchise current, and as a consequence of that, the City of Bellingham had the opportunity as a controller of property to enter into an agreement with Olympic which required it to do very specific things, very specific things that will ensure the safety and the protection of the people of Bellingham, things like hydrostatically testing its pipeline, a thorough review of its computer monitoring system, a review of the placement, location and operation of all the valves, a provision for a leak detection system, provision for a thorough and adequate staffing and training program, and comprehensive analysis of the entire Olympic system to ensure that all aspects of its operation including management, training and operating procedures are sufficient to ensure that it will operate in a safe fashion. We were able to do this simply, because they failed to renew their franchise a few years ago.

Senator GORTON. How long is the franchise period?

Mayor ASMUNDSON. The franchise period was 30 years, and it had expired. The franchise was entered into in 1965, and it expired in 1995 and just through lack of continuity, it was not renewed. So we had this window of opportunity to engage in very sophisticated safety protection for the people of Bellingham.

Ironically, we did that in 3 months. We achieved for the people of Bellingham what I think is a blueprint for the other communities in Washington and should be a model for pipeline safety for this country. But the ironic question I have to ask is, with no background or history in dealing with fuel pipeline operations, the City of Bellingham in 3 months was able to develop a comprehensive pipeline safety program that truly meets the needs of our community both now and into the future—why is it that after receiving millions upon millions of dollars in Federal funds, having an expert staff and decades to accomplish this, the Federal Office of Pipeline Safety has been unable to do so? In 3 months we came up with a package that will protect citizens.

The Office of Pipeline Safety in response to this accident has been very attentive. It has adopted many, many of the things that we achieved through our pipeline safety agreement with Olympic Pipe Line Company, but I think what the citizens demand is not a Federal agency that is very capable of coming up with corrective action orders after an injury, after a death, or after an environmental disaster, but rather an agency that is oriented toward prevention and looks to the future.

The Office of Pipeline Safety must be made accountable. The state legislature has taken the steps it needs to take in order for the State of Washington to be a strong partner in protecting our citizens, but the effectiveness of that legislation does not lie with the Governor as to whether he'll sign it or not. The effectiveness lies with the Federal Government and whether or not adequate authority will be given to the states to make a meaningful difference to protect our citizens in our community and our environment.

I really must thank Senator Murray and Senator Gorton. I must thank you both for the incredible hard work that you have done. I need to thank Jean for her hard work. I need to thank Dale and the rest of your staff for the hard work that they've done on this. The people of Bellingham and Washington State deserve to know how hard you have worked on this issue and how important that is to us, and I personally thank you for your commitment to making a difference, to seizing this opportunity to truly make a difference and change the status quo.

Stephen Tsiorvas, Wade King and Liam Wood did not deserve to die on June 10th, 1999. You know that, and it's my hope that the memory of these wonderful boys and their needless sacrifice will encourage you and all the members of the senate and Congress to ensure that no other parents, no other elected officials, no other friends must sit before another committee of the U.S. Senate at any time in the future and repeat the tales that you've heard today.

Thank you for the opportunity to be here.

[The prepared statement of Mayor Asmundson follows:]

PREPARED STATEMENT OF MARK ASMUNDSON, MAYOR, BELLINGHAM, WASHINGTON

My thanks to you, Senator Gorton, and to Chairman McCain and other members of the Committee for conducting the field hearing today. I appreciate the opportunity to testify before this Committee. I, and the citizens of Bellingham appreciate the opportunity to discuss the important subject of pipeline safety and to highlight the inadequacies of the current methods of oversight of the safety of the interstate fuel pipeline network in the United States.

In Bellingham, we have experienced, in a dramatic, tragic, and profound way, the failure of the current system of ensuring pipeline safety in America.

As you know, on June 10, 1999, the Olympic pipeline, which passes through Bellingham, ruptured, spilling one quarter of a million gallons of gasoline into a park and creek in the middle of our city. The gasoline vapors ignited and two boys and a young man were killed as a result. The city park was severely impacted and the salmon-bearing stream was effectively sterilized for one and one-half miles of its length. But for the inadvertent ignition by two of the boys, the gasoline would undoubtedly have proceeded into and through downtown Bellingham into Bellingham Bay, the result of which would have been even further devastation to our community.

It goes without saying that the people of Bellingham are concerned about the safety of the Olympic pipeline and the safety of pipelines throughout our country. The Olympic pipeline travels through twenty-one cities in the State of Washington. The combined population of these cities is nearly 800,000 people. A significant portion of this pipeline passes through highly urbanized areas, as well as environmentally sensitive areas including lakes, salmon-bearing streams, aquifers and critical wetland habitats.

In Bellingham, the Olympic pipeline travels through many neighborhoods, near schools, across city parks, and traverses three salmon-bearing streams.

Following the disaster, I began a period of intense involvement in understanding the fuel pipeline industry, the regulatory regime, and the circumstances that could have led up to the kind of event that occurred in Bellingham on June 10, 1999. I discovered that the Bellingham event was not unique. Based on the history of the pipeline industry and the spills that have occurred over recent years, I have concluded that this event could have been expected.

Given the current state of affairs involving interstate fuel pipelines in America, the issue is not *will* pipelines leak, but *when* will they leak, *where* will the leaks occur, and what kind of harm, great or small, will result from the leak or rupture? Over the course of the last two decades, there is a history of disaster after disaster involving interstate pipelines. Many of these have resulted in serious injuries or death. All of them have resulted in serious environmental damage. (Attachment "A" highlights several significant leaks.)

Having looked at the history of pipeline accidents in America and finding that the track record of safety for pipelines is truly alarming, the question I had to ask myself is: Why? How can this continue to happen?

While the ongoing occurrence of accidents such as the one that occurred in Bellingham causes alarm, the answer to the question of why these continue to occur is even more alarming.

The federal government has preempted regulation of pipeline safety. It is clearly within the power and purview of the federal government to do so. The alarming fact is that while the federal government has prevented states and localities from engaging in safety and environmental protection-oriented regulation of pipelines, it has not come remotely close to ensuring that pipelines will be operated in a safe fashion. In fact, in reviewing the many recent comments of Mr. Hall, the chairman of the National Transportation Safety Board, it is apparent that the OPS has consistently failed to take steps reasonably calculated to result in safer pipelines in our communities.

Why is it that the OPS has done a poor job of making pipeline operations safe? In part, it is because of limited resources. It is true that the OPS has not been adequately funded, particularly until the most recent past. However, as a result of my observations and review of the record, I have concluded that the fundamental problem is a lack of *will* on the part of the OPS. Rather than aggressively and effectively pursuing public safety, the OPS seems intent on ensuring that it takes no action without the agreement and concurrence of the industry it is charged with regulating.

I have also discovered that the OPS has failed to comply with the explicit mandates of Congress with regard to safety and environmental protection. For example, Congress has mandated that the OPS adopt regulations for unusually sensitive areas and adopt regulations regarding use of internal inspection devices. The deadlines for completing these actions passed years ago. The OPS has simply failed to accomplish these mandates. That is not to say that the OPS did not undertake these mandates, but it is curious to note that whenever agreement with the industry could not be achieved, the regulatory process effectively ground to a halt. Even as you hear this testimony today, and witness for yourselves our tragedy, the regulatory process has not produced the safety regulations required by Act of Congress.

As I stated earlier, the Bellingham incident is not unique. Other communities have experienced similar tragedies, and as a consequence, like Bellingham, and like the State of Washington, these communities have been awakened to the inadequacies of the federal regulatory environment. Other states have sought to improve the safety of their citizens through establishing comprehensive programs of pipeline safety within their states. The full effectiveness of these efforts, however, has never been realized, and the safety potential that could have resulted from these proactive actions by states has never been achieved because of the unwillingness of the federal government to release its stranglehold on regulatory authority over the pipeline industry even when that federal regulatory authority stands unused and unfulfilled. Consequently, states like Minnesota and California, which have established very thorough programs for pipeline safety, have never been able to achieve their full potential because of the unwillingness of the federal government to allow states to have a meaningful role in ensuring the safety of their own citizens and the desire of the OPS to maintain complete control over every aspect of the industry. This desire for complete control by the OPS is demonstrated by its determination to withdraw from the agency arrangements that it has in place with four states authorized in the past to inspect, but not regulate federal interstate pipelines within their boundaries.

The City of Bellingham, with many other cities and counties in the state, has been working for passage of state legislation authorizing much more extensive protection of our citizens related to pipeline safety. The question as to whether or not this effort will be successful is not, however, in the hands of the state legislature or the governor, but in the hands of the federal government. Unless Congress is willing to allow states to have a meaningful role in ensuring the protection of their citizens through an active program of regulation of interstate pipelines, this effort, like the efforts in California and Minnesota, simply will not accomplish needed safety and environmental protection.

The rules under which this industry operates simply must change. In order for that to occur, changes are required by the federal government. Since the federal government has shown its unwillingness to regulate effectively this industry and provide the protection that our citizens and environment need, deserve and demand, it is the responsibility of the federal government to allow states to protect their citizens. This is not to say that the federal government should abandon the regulation of pipelines. I would be the last to argue for such a step. Clearly, the federal government must establish a minimum level of safety that must be followed by any interstate pipeline operator. However, the federal government's legitimate concern about interference with interstate commerce need not prevent states from providing for the safety of their citizens. A partnership can be established between the federal government and states whereby states can protect their citizens without interfering with interstate commerce, particularly in cases where the states have expertise or where the federal government could have acted, but the federal regulatory agency has chosen not to act.

It is for this reason that I heartily support Senate Bill 2004 offered by Senator Murray and cosponsored by Senator Gorton. I believe that this bill, together with the similar effort in the House of Representatives, as set forth in HB 3558, introduced by Representative Metcalf, will accomplish the changes necessary to ensure that not only do we have a strong, viable fuel distribution system in America, but we have a strong, viable fuel distribution in America that can and will be operated in such a fashion that it does not endanger our environment; does not kill our children; and allows our citizens living near pipelines to go to sleep at night without anxiety about their personal safety.

My testimony thus far consists of an overview of the results of my quite extensive examination of pipeline safety in America, the Federal OPS, and the attempts by certain states to improve the safety of their citizens. I would now like to discuss what pipeline safety means in real, practical, human terms.

Fundamentally, pipeline safety is not about legislation, and it is not about federal bureaucracies. Pipeline safety is about people and the environment. When a pipeline ruptures, it doesn't rupture in a committee hearing, and it doesn't rupture in an agency office; it ruptures in a neighborhood, or a park, or a wetland, or over a stream, or in a farmer's field.

The effects of the rupture are not abstract or theoretical, they are real. They are practical. In Bellingham's case, it meant attending the funerals of three wonderful boys in one week, along with a community of thousands of mourners. It meant the disruption of the water supply to one-quarter of the citizens of our community, and restrictions on water use, while alternate facilities were being developed to provide an adequate water supply. It meant the indefinite delay of salmon restoration projects on a salmon-bearing urban stream. It meant the devastation of one and one-half miles of wildlife habitat along a stream corridor in the midst of one of the older and best-established parks in our community. It meant children afraid that their neighborhoods may blow up. It meant anguish, questioning, grieving, and, in some cases, a totally unsatisfying search for rational answers to the question, why?

In Bellingham's case, it means great frustration on the part of local elected officials like myself. When asked by citizens, "what will you do to make sure that we are safe?" the answer, "I will do the best I can, but my hands are tied by federal law" does not satisfy. In addition to the mayors here today, I wish it were possible for there to be sitting at the table with me, Mayor George Spadoro of Edison, NJ, City Councilman Robert Mann of North Blenheim, NY, and Mayor Bill Greenup of Fredericksburg, VA. They are just a few of the mayors and other local elected officials we have reached out to who could tell you about the enormity of the disaster visited on their communities by pipeline accidents over the years. I believe they would echo my comments.

While the pipeline disaster in Bellingham was not unique, in a very important way, the Bellingham situation *is* unique. For reasons unknown, in 1995, the Olympic Pipe Line Company ("Olympic") allowed its franchise (its right to cross city property), to expire. As a consequence, the City of Bellingham was in a strong bargaining position to require that very explicit safety measures be taken by Olympic as a condition for permission to utilize our property. As a result of this quirk, we have been able to take steps that we believe will ensure the protection of Bellingham's citizens. These steps included requiring: Hydrostatic testing of the pipeline; a thorough review of the computer monitoring system; a review of the placement, location, and operation of all valves; the provision of adequate leak detection systems; provision for thorough and adequate staffing and training; and a comprehensive analysis of the entire Olympic pipeline system to ensure that all aspects of the system, including its management and operating procedures, are sufficient to ensure that it will operate in a safe fashion. Remember, though, we were able to do this because of Olympic's failure to renew its franchise a few years before this tragic accident.

The question, however, is "why is it necessary for a community to rely on a glitch, a happenstance, a mistake by the pipeline company in the maintenance of its franchise, for us to have the ability to protect our citizens?" It shouldn't be.

With virtually no background or history in dealing with fuel pipeline operations, the City of Bellingham, in a period of approximately three months, was able to develop a *comprehensive* pipeline safety program that meets the needs of our community.

Why is it after receiving millions upon millions in federal funds, having an expert staff and decades to accomplish the same, the OPS has not been able to do so? Honorable members of the Committee, the reason that this has not occurred is because the *will* to do so has not been present.

We have, through our actions, provided for the safety of our citizens. We are confident that Olympic's pipeline, should it restart, will be safe because of the actions we have taken in our pipeline safety plan. What we have achieved in such a short time is something that the citizens of every community in this state and the citizens throughout this country, are entitled to receive. They will not receive this level of protection unless action is taken to change the status quo. The OPS has not adequately served the public interest. Congress needs to take steps to protect our citizens and our environment. The OPS must be made accountable. It must be required to fulfill its mandate. Senator Murray's legislation creates an environment which will result in an accountable agency and a responsive system of pipeline safety.

Steven Tsiorvas, Wade King, and Liam Wood did not deserve to die on June 10, 1999. May the memory of these wonderful boys and this needless sacrifice encourage you to take steps that will ensure that no other parents must sit before another Committee of the United States Senate at some time in the future and repeat the tales that you have heard today.

ATTACHMENT "A"

Examples of Major Pipeline Accidents (1980–1999)

Accidents

(1) *Fredericksburg, VA* 1980 (and again in 1989)

330,000 gallons of aviation fuel entered the city water supply, and the Rappahannock River, shut down the water treatment plant, a state of emergency was declared, and businesses and residents hauled water for a week.

Causes: Pipe damage upon installation, subsequent undetected corrosion, operator error, insufficient valve placement.

(2) *Moundsview, MN* 7/8/86

An 8-inch gasoline pipeline burst, gasoline flowed along neighborhood streets until it was ignited, killing 2 people who burned to death and injuring 7.

Causes: Failure to correct known defects, inadequate pipe specifications, inadequate operator training including delay in responding.

(3) *Flathead Indian Reservation, MT* 1986–1993

Seventy-one leaks and three major spills of gasoline, aircraft fuel, and diesel (including 163,000 gallons into a creek) over this period resulted in the Flatheads refusing to renew Yellowstone's franchise and move the line off of the reservation.

Causes: Inattention and failure to correct defects.

(4) *North Blenheim, NY* 3/13/90

A liquid natural gas pipeline burst sending 100,000 gallons of product flowing down into the town—enough to engulf the entire town. Residents noticed a "heavy fog" on their windshields, until one called and notified a company employee. Two people were killed and seven injured.

Causes: Negligent maintenance procedures resulting in cracks in the pipe which were undetected; operator error; insufficient remotely operated valves and check valves.

(5) *Reston (Herndon), VA* 3/28/93

Pipe burst sending a geyser of diesel fuel (407,000 gallons) into the storm sewer and eventually into a tributary of the Potomac River. (Could have been gasoline or jet fuel.) Significant environmental damage ($1 million clean-up).

Causes: Third-party damage causing corrosion which went undetected for a long period.

(6) *Edison, NJ* 3/23/94

Natural gas transmission line burst and exploded. 1500 residents evacuated and $25 million damage. Injuries included minor burns and cuts from broken glass.

Causes: Line hadn't been "pigged" since 1986, but it had deteriorated; no remote automatic valves; pipe manufacturing standards lax; no extra measures for highly populated areas.

(7) *Allentown, PA* 6/9/94

Natural gas pipe burst and product flowed underground into the basement of an 8-story retirement home, where it migrated through vents into other floors and was eventually ignited. One death and 55 injuries.

Causes: Company employee (backhoe operator) error; no excess flow valves which had been recommended by NTSB since 1972; insufficient public awareness.

(8) *Reedy River, SC* 6/26/96

Fuel oil pipeline crossing the river burst resulting in a $20 million clean-up effort.

Causes: Pipeline corrosion not responded to soon enough; computer malfunction; employee error; inadequate leak detection.

(9) *Lively, TX* 8/8/96

Liquid natural gas pipeline burst, killing two men who accidentally ignited it.

Causes: Inadequate corrosion protection.

(10) *Murphreesboro, TN* 11/5/96

84,000 gallons of diesel fuel (could have been gasoline) and $5.7 million damage. *Causes:* Corrosion; operator error—3½ hours before detection.

(11) *San Juan, PR* 11/21/96

Thirty-three people killed when a liquid natural gas line exploded. *Causes:* Employee negligence in responding to a leak which had been ongoing for a week.

(12) *Bellingham, WA* 6/10/99

Notes

None of these accidents were the result of "third party damage" with the exception of the Reston incident.

All of them could have been prevented—if safety recommendations had been acted upon.

The common causes of pipeline accidents are:

- anomalies in the pipe not detected or not acted upon
- operator inattention or error
- computer system malfunction
- shut-off capability insufficient or improperly deployed
- leak detection insufficient

On average 6 million gallons spilled each year; 8 million last year.

Since 1996, 54 accidents investigated by NTSB, resulting in 209 recommendations.

Since 1986, 39 deaths/189 injuries (natural gas); 35 deaths/247 injuries (liquid products).

Pipeline companies involved: Colonial, Yellowstone, Koch, Olympic, Williams, Texas Eastern, and others.

Koch just fined $30 million for 300 separate spills of 3 million gallons in six states between 1990 and 1997; leak detection system: wait until it breaks. (See attached article.*)

Further resource: Battelle Labs' "Causes of Pipeline Incidents, Effect of the Aging Infrastructure on Incidents and Areas of Technology Development," Robert J. Eiber (1994), delivered at an OPS summit on pipeline safety.

Senator GORTON. Thank you, Mayor.

Mayor Tanner?

STATEMENT OF JESSE TANNER, MAYOR, RENTON, WASHINGTON

Mayor TANNER. Senator Gorton, Senator Murray, thank you for the opportunity to testify on the petroleum pipeline safety, and the provisions we feel are necessary in pipeline safety reauthorization legislation.

Renton serves as the headquarters for the Olympic Pipe Line Company and many miles of petroleum pipeline run through Renton. There have been two major Olympic Pipe Line production leaks in Renton. In 1986 a leak at a blocked valve in the Maplewood residential neighborhood resulted in 80,000 gallons of gasoline entering the ground water, seeping into the Cedar River which is home to the largest sockeye salmon run in the lower 48 states, and causing explosive levels of fumes in five homes. These homes where evacuated for a week. Remnants of petroleum contamination still exist within a 1,500 foot long ground water plume. The only reason that ground water plume has not leached into our aquifer, our drinking water aquifer, is there's an aquitard layer that prevents it. It's still there, and the possibility still exists that will invade our aquifer.

*The information referred to was not available at the time this hearing went to press.

In August 1999, 3,500 gallons of petroleum product from a broken pump at Olympic Pipe Line's Lind Avenue control center escaped to the ground.

Internal smart pig testing of the pipelines in 1996 through 1997 revealed over 270 pipeline flaws, 15 of which are in Renton, and they're shown on that large map there. There's a small map attached to the material which I provided to the committee.

Even though Renton's franchise contract with Olympic Pipe Line requires that such information be submitted to Renton, this data was not received until October 1999. In fact, Renton encountered a great deal of difficulty in obtaining this data.

After the Bellingham incident, Olympic Pipe Line refused to provide the data on legal grounds. When Renton cited the franchise requirements, Olympic Pipe Line continued to delay until Renton issued an ultimatum that the information be provided within 30 dates or else the franchise agreement would be terminated, and Renton would request the pipelines to be removed from the city.

I don't know how effective that would have been, but anyway, that's what happened. On the 30th day Olympic provided the smart pig testing data, but it was in a format that could only be understood by Olympic Pipe Line employees. The stationing data for the links along the pipeline were not provided.

Finally enough information was provided to produce the attached map showing the pipeline flaws in Renton. The map that has been submitted to you shows the location of these pipeline flaws, and of the 15 pipeline flaws in Renton, three involve pipeline metal loss of 50 percent. One of these major pipeline flaws is within 300 feet of Talbot Hill Elementary School, and the other two are located over Renton's sole source drinking water aquifer.

The Olympic Pipe Line Company has not repaired or even inspected these flaws. They feel that they are not required to by the American Society of Mechanical Engineers pipeline regulations which they use, which allows pipelines to be operated with up to 80 percent metal loss under certain circumstances.

Renton disagrees with the Olympic Pipe Line's reading of the ASME requirements. These standards call for pipeline sections with gouges and grooves deeper than 12 and a half percent of the pipeline wall thickness to be removed or repaired. Pipeline operation with wall metal loss with up to 80 percent is only allowed if metal loss is due to corrosion pitting, and even then the corroded area must be recoated.

Since the Olympic Pipe Line company did not visually inspect the pipeline flaws in Renton, they cannot be sure of what caused the flaws, corrosion pitting or external damage. They have not met the requirement of either provision nor has the Office of Pipeline Safety required them to do so.

Olympic Pipe Line Company objects to hydrostatic pressure testing the whole pipeline because of concerns about the test damaging the pipeline. We're not aware of any empiric data that supports this concern. If the company is worried about pressurized water damaging the pipelines, why do we not see a corresponding worry about transient pipe pressure surges damaging the pipeline when it is filled with gasoline?

Congressman Jay Inslee has obtained information indicating that even high frequency electric resistance welded pipe has a history of pipeline seam failure. The Office of Pipeline Safety has required hydrostatic pressure testing only for low frequency electric resistance welded pipe. This information from our Representative Inslee indicates that no distinction should be made. The entire line should be hydrostatically pressure tested.

Renton is concerned about leak detection. As shown by the 1986 Maplewood leak which was 80,000 gallons occurring over perhaps months of time, smaller but extremely dangerous leaks cannot be detected by Olympic Pipe Line Company's pressure sensors. Federal regulations should require improved leak detection technology to be implemented particularly in population centers and sensitive areas.

We also believe that states should be allowed to require use of improved technology that would enable leaks to be quickly detected and located.

Renton strongly supports S. 2004 and H.R. 3558 currently before Congress, and we think that each one of those bills has some provisions that the other does not, so we would suggest that there is, that the provisions of both bills be combined.

Renton is particularly supportive of requiring hydrostatic pressure testing every 5 years and the requirement for improved leak detection technology to be implemented.

Renton also supports the other provisions of both bills. I would like to close by requesting congressional support in persuading the Office of Pipeline Safety to issue a corrective action notice to Olympic Pipe Line Company to require the entire 400 mile pipeline system to be hydrostatically pressure tested. This is a very necessary stop gap measure that will protect thousands of people exposed to the pipeline until more comprehensive safety regulations can be put into place. Thank you.

[The prepared statement of Mayor Tanner follows:]

PREPARED STATEMENT OF JESSE TANNER, MAYOR, RENTON, WASHINGTON

Thank you Senator Gorton for the opportunity to provide testimony before you today. The citizens of Renton, Washington, have a special interest in the safety of hazardous liquid pipelines in that Renton serves as the headquarters of the Olympic Pipe Line Company, and many miles of petroleum pipelines run through our city.

I have been invited to speak to you about our recommendations on options the Congress could consider to improve the transportation of hazardous liquids through pipelines in the context of the Committee's pipeline safety reauthorization legislation later this year. I am honored to do so, and I believe that our experience and history with this subject qualify us to present you with facts and perspectives that will be useful to you in your deliberations on the pipeline safety reauthorization legislation.

Olympic Pipe Line Company and Renton

In addition to having their headquarters in Renton, the Olympic Pipe Line Company owns and operates a series of pipelines within the city. The parallel 16-inch and 20-inch lines that run north-south through the city were installed in 1965 and 1973 respectively. The joint capacity of these lines exceeds 14.7 million gallons of gasoline, diesel fuel and jet fuel per day. These large lines carry petroleum product to the Lind Avenue pump station in Renton, where fuel is pumped to Portland, Oregon through a 14-inch line installed in 1965, and to Harbor Island, Seattle and

Sea-Tac Airport through two 12-inch lines installed around 1970. The attached map shows the routing of these pipelines in Renton.*

The 16-inch pipeline (which failed on June 10, 1999 in Bellingham) is made of .312-inch wall steel pipe, and the 20-inch line uses .25-inch pipe. These petroleum pipelines run through residential neighborhoods, past schools, and over Renton's drinking water supply aquifer along much of their length.

Olympic Pipe Line's history in Renton is somewhat checkered. There have been two major fuel leaks. In early October 1986 an estimated 80,000 gallons of mixed gasoline, diesel and jet fuel were discovered to have leaked into the Maplewood residential neighborhood. The fuel product was released gradually, perhaps over a period of weeks or months, at the location of a block valve. The leak was not detected by the Olympic Pipe Line Company, but rather by citizens who noticed an iridescent plume spreading into the nearby Cedar River. The Cedar River, incidentally, is home to the largest sockeye salmon run in the lower 48 states. At about the same time, some residents of Maplewood noticed gasoline fumes in their basements, and the Renton Fire Department was called to the scene. Explosive levels of fuel vapors in basements caused five families to be evacuated from their homes for a week. Investigation determined that the fuel had contaminated a 1,500 foot-long plume eight to twenty-three feet underground. The presence of an impervious aquitard layer under the Cedar River caused the contaminant to be released into the Cedar River rather than percolating downward to contaminate Renton's drinking water supply. Olympic Pipe Line Company provided an 18-month remediation program consisting of ground water pumping, floating petroleum recovery and soil vapor extraction. In 1998 the Washington State Health Department initiated a round of monitoring and testing at Maplewood which once more revealed elevated petroleum hydrocarbons in the ground water. These elevated readings turned out to be contaminants left over from the 1986 event rather than from a new leak. At that time the Olympic Pipe Line Company indicated the intent to leave the petroleum product in the ground, but after the application of significant pressure by Renton, determined to remove the product by use of the air sparging process.

The second major fuel leak occurred on August 29, 1999. This spill took place at the Olympic Pipe Line Company's Renton Lind Avenue headquarters as the result of a transfer pump that had broken leaking product onto the ground. Approximately 3,500 gallons of fuel escaped over an approximate 40-minute period before the leak was discovered by company employees. It was over an hour before the company contacted the Renton Fire Department.

In 1996–1997 the Olympic Pipe Line Company conducted an internal "smart pig" test throughout their pipelines to determine the condition of these aging lines. Over 270 "anomalies" or flaws were found at that time. Although at least 15 of these flaws are located in Renton, and in spite of the fact that the Olympic Pipe Line Company had entered into a legal agreement (franchise) to turn over all test results to the city, the City of Renton did not receive this information until October, 1999. Furthermore, the information was only received after repeated requests that included an ultimatum that we would terminate the franchise agreement and request that the pipelines be removed from Renton. Now that we have the test results, and have mapped the flaws, we have significant concerns about the safety of the pipelines. I will discuss these concerns later in this testimony.

Renton's Concerns

We are very aware of, and nervous about, the tragedy that occurred in Bellingham on June 10, 1999. We think that it is possible that such an event could happen again. In fact the evidence, and lack of regulatory oversight, tends to make us think that it is likely that such an incident will happen again, if not in Renton, elsewhere along the pipeline.

What is this evidence? First, I will speak of the general considerations. These are aging pipelines—some sections are up to 35 years old. They are high pressure and high volume lines, with a maximum operating pressure up to 1,400 pounds per square inch. The pipelines are metallic, and are therefore subject to corrosion. The product being carried consists of highly incendiary, explosive grades of refined petroleum. The pipelines run through residential neighborhoods and schoolyards, beneath environmentally sensitive areas, and across salmon-bearing streams. Oversight is provided by a severely understaffed and underfunded federal regulatory agency which, until recently, did not even have an inspector based in the state of Washington. The current federal regulations call for little or no mandatory pipeline or system testing, and do not provide means for assuring safety of the operation. There appears to be no requirement for public disclosure, little oversight on operator train-

---

*The information referred to has been retained in the Committee's files.

ing, and no requirement for cooperation, or even communication, with local emergency response agencies. State and local governments are preempted from involvement in the regulation of this industry, and regulation is only minimally provided by the federal government. Even without more specific information, these ingredients seem to be a recipe for disaster.

However, we do have more specific information, and that information is chilling. The attached map shows the route of these pipelines through the City of Renton. The blue and pink areas of the map depict Renton's drinking water aquifer protection areas. These areas occur over the top of our sole source potable water supply, so that any petroleum leaks in these areas could have catastrophic consequences to our drinking water. The green lines on the map represent the Olympic Pipe Line petroleum pipes, the blue boxes are schools, and the 15 flaws are flagged out along the routes of the pipes. A key at the lower right hand corner of the map shows what the numeric information in the callouts means. Five of the pipeline flaw callout boxes are red—these indicate the more serious flaws (between 29% and 57% of metal loss in the pipeline wall!). Metal loss indicates the percentage of metal that is missing in the pipeline wall. The most serious pipeline flaw in the City, with 57% metal loss, is located within 300 feet of Talbot Hill Elementary School. Two pipeline flaws with roughly 50% metal loss are located over our drinking water supply. Most of these pipeline flaws are in heavily populated residential neighborhoods.

Back in 1996–1997 when the Olympic Pipe Line Company acquired this information through "smart pigging," they were not sufficiently concerned to perform any further investigation. They did not provide this information to local government, to school districts, or to residents. It remains the Olympic Pipe Line Company's position today that no further action needs to be taken to address these pipeline flaws. They cite the governing standards, ASME B31.4, as allowing corrosion pitting of the pipeline wall up to 80% loss of wall thickness before replacement is required. However, paragraph 451.6.2(a)(1) of this standard states that gouges and grooves having a depth greater than 12½% of the nominal wall thickness shall be removed or repaired. How does Olympic Pipe Line Company know, without visual inspection, whether the areas of metal loss are due to corrosion or to gouges or grooves? And even if the metal loss were due to corrosion, the ASME B31 Supplemental Manual for Determining the Remaining Strength of Corroded Pipelines states, "in all cases where the corroded region is to be left in service, measures should be taken to arrest further corrosion. Such measures should include coating the corroded region and, if indicated, increasing the cathodic protection level." To our knowledge, the Olympic Pipe Line Company has not ascertained the cause of the pipeline flaws in Renton, has not taken measures to arrest further corrosion, nor has the Office of Pipeline Safety required that this information be ascertained or that any remedial measures be taken.

There is only one way that we are aware of to assure that these aging, pitted pipelines can sustain the required test pressure without failing, and that is to hydrostatically pressure test the pipelines. The Olympic Pipe Line Company refuses to do this, and unaccountably, the Office of Pipeline Safety has declined to force them to do so through a Corrective Action Order. The Olympic Pipe Line Company proposes to undertake another "smart pig" internal inspection of the pipelines instead. Renton takes no issue with performing another round of internal testing. However, the results of such testing cannot be correlated to pipeline strength. Until a test is undertaken that can demonstrate the current strength of the pipeline, no one can say what pressure or operating conditions the pipeline will support. This can only be determined by a hydrostatic pressure test.

One of the objections that the Olympic Pipe Line Company has raised regarding hydrostatic pressure testing of the pipelines is the potential that such testing could damage the pipelines. However, the company to our knowledge has not presented scientific evidence to support this claim. Hydrostatic pressure testing of pipelines is a standard test procedure that has been in use for decades to demonstrate that pipelines are capable of sustaining their rated test pressures. If performed correctly, such testing should not result in damage to sound portions of the pipeline. It is true that weakened portions of the pipeline could fail, as was the case with the hydrostatic testing in Bellingham. This is the very purpose of the test: to identify weakened, dangerous portions of the pipeline so that these sections can be replaced to prevent future catastrophes such as happened in Bellingham. Our question for the Olympic Pipe Line Company is: if you are so worried about test pressures damaging the pipelines when the pipe is carrying water, why do you not appear to be worried about transient surge pressures that also exceed the operating pressure causing damage when the pipe is carrying petroleum products?

I would like to mention one other issue concerning strength of pipelines. The Office of Pipeline Safety's current Corrective Action Orders require hydrostatic testing only for sections of the Olympic Pipe Line that are low frequency electric resistance welded (ERW) pipe. These sections occur mainly in Whatcom County, in the Bellingham area. However, Congressman Jay Inslee has investigated the failure history of high frequency and low frequency ERW pipe. He has found information from the Office of Pipeline Safety web site archives that compares the number of failures of electric-resistance longitudinal welded pipes manufactured by U.S. Steel Corporation, and pipe manufactured by Lone Star Steel from 1970 to mid 1984 on gas pipelines. This information seems to contradict the assertion that U.S. Steel does not have a seam failure history for high-frequency ERW manufactured pipe. It is my understanding that the majority of the pipeline is made of U.S. Steel and other brands of high-frequency ERW manufactured pipe. A table showing Congressman Inslee's findings is attached. This information calls into question the Office of Pipeline Safety's distinction between the reliability of Lonestar and U.S. Steel pipe, and supports the argument that the entire pipeline should be hydrostatically tested.

Renton is also concerned about leak detection. Currently the Olympic Pipe Line Company detects leaks by internal pressure loss. As demonstrated by the Renton Maplewood leak, more gradual leaks, that can also be devastating to safety and the environment, cannot be detected by the Olympic Pipe Line Company's current technology. We feel that the Federal regulations should require improved leak detection technology, particularly in population centers and sensitive areas.

Renton's Recommendations

The City of Renton supports both the Pipeline Safety Act of 2000 (S. 2004) and the Safe Pipelines Act of 2000 (H.R. 3558). We particularly support the provision of H.R. 3558 that requires hydrostatic testing of all facilities once every 5 years, and the provision of S. 2004 requiring the use of equipment to detect and locate leaks. We support improved certification and testing of operators, improved corrosion testing, better notification of spills (particularly of the local agencies which are charged with providing emergency response), and delegation of authority to states. Providing additional funding to the Office of Pipeline Safety also seems like a sound provision. This agency does not appear to us to be equipped to handle the challenges of regulating petroleum pipeline safety in the face of growing distribution systems and aging infrastructure. Providing additional funding to this agency would seem to be particularly essential if more regulatory authority is not delegated to the states.

Finally, I would like to take this opportunity to make an appeal to you to take a step that would help the citizens of Renton and others living along the route of the pipelines sleep better at night. This appeal is to contact Kelly Coyner, Administrator of the Office of Pipeline Safety, and request that a Corrective Action Order be issued to the Olympic Pipe Line Company requiring hydrostatic testing of the entire length of their pipelines. By influencing the Office of Pipeline Safety to do so, defective sections of the pipeline could be detected and repaired in the near term, which would significantly reduce the risk of catastrophic failure over the next several years.

Once again, I wish to thank you for inviting me to participate in a process that could provide much needed protection to the citizens of Renton as well as to other residents throughout the country whose proximity to hazardous liquid pipelines exposes them to risks that are not currently sufficiently regulated.

U.S. Steel Corp and Lone Star Steel
Seam Failures

US Steel Corp

	Incident ID	State	Year	Leak Origin	Cause of Leak	Type of Steel
1	3770001	TN	1978	LONGITUDINAL WELD	MATERIAL FAILURE	ELECTRIC-RESISTANCE WELDED
2	2790021	TX	1979	LONGITUDINAL WELD	MATERIAL FAILURE	ELECTRIC-RESISTANCE WELDED
3	3790073	TX	1978	LONGITUDINAL WELD	MATERIAL FAILURE	ELECTRIC-RESISTANCE WELDED
4	3700328	OH	1970	LONGITUDINAL WELD	MATERIAL FAILURE	ELECTRIC-RESISTANCE WELDED
5	3700020	TX	1970	LONGITUDINAL WELD	MATERIAL FAILURE	ELECTRIC-RESISTANCE WELDED
6	3700583	OK	1970	LONGITUDINAL WELD	MATERIAL FAILURE	ELECTRIC-RESISTANCE WELDED
7	2802011	OH	1980	LONGITUDINAL WELD	MATERIAL FAILURE	ELECTRIC-RESISTANCE WELDED
8	3700402	CA	1970	LONGITUDINAL WELD	MATERIAL FAILURE	ELECTRIC-RESISTANCE WELDED
9	2820719	NM	1982	LONGITUDINAL WELD	MATERIAL FAILURE	ELECTRIC-RESISTANCE WELDED
10	2820720	NM	1982	LONGITUDINAL WELD	MATERIAL FAILURE	ELECTRIC-RESISTANCE WELDED
11	2820728	NM	1982	LONGITUDINAL WELD	MATERIAL FAILURE	ELECTRIC-RESISTANCE WELDED
12	2820729	NM	1982	LONGITUDINAL WELD	MATERIAL FAILURE	ELECTRIC-RESISTANCE WELDED
13	2820909	NM	1982	LONGITUDINAL WELD	MATERIAL FAILURE	ELECTRIC-RESISTANCE WELDED
14	2820910	NM	1982	LONGITUDINAL WELD	MATERIAL FAILURE	ELECTRIC-RESISTANCE WELDED
15	3700302	AZ	1970	LONGITUDINAL WELD	MATERIAL FAILURE	ELECTRIC-RESISTANCE WELDED
16	3700303	AZ	1970	LONGITUDINAL WELD	MATERIAL FAILURE	ELECTRIC-RESISTANCE WELDED
17	3700304	AZ	1970	LONGITUDINAL WELD	MATERIAL FAILURE	ELECTRIC-RESISTANCE WELDED
18	3700449	WY	1970	LONGITUDINAL WELD	MATERIAL FAILURE	ELECTRIC-RESISTANCE WELDED
19	3720249	WY	1972	LONGITUDINAL WELD	MATERIAL FAILURE	ELECTRIC-RESISTANCE WELDED
20	3760004	NM	1976	LONGITUDINAL WELD	MATERIAL FAILURE	ELECTRIC-RESISTANCE WELDED
21	3760006	CO	1976	LONGITUDINAL WELD	MATERIAL FAILURE	ELECTRIC-RESISTANCE WELDED
22	3760007	CO	1976	LONGITUDINAL WELD	MATERIAL FAILURE	ELECTRIC-RESISTANCE WELDED
23	3720111	CO	1972	LONGITUDINAL WELD	MATERIAL FAILURE	ELECTRIC-RESISTANCE WELDED
24	3720112	WY	1972	LONGITUDINAL WELD	MATERIAL FAILURE	ELECTRIC-RESISTANCE WELDED
25	3720243	WY	1972	LONGITUDINAL WELD	MATERIAL FAILURE	ELECTRIC-RESISTANCE WELDED
26	3720244	WY	1972	LONGITUDINAL WELD	MATERIAL FAILURE	ELECTRIC-RESISTANCE WELDED
27	3720245	WY	1972	LONGITUDINAL WELD	MATERIAL FAILURE	ELECTRIC-RESISTANCE WELDED
28	3720248	WY	1972	LONGITUDINAL WELD	MATERIAL FAILURE	ELECTRIC-RESISTANCE WELDED
29	3730122	CA	1972	LONGITUDINAL WELD	MATERIAL FAILURE	ELECTRIC-RESISTANCE WELDED
30	3730123	CA	1972	LONGITUDINAL WELD	MATERIAL FAILURE	ELECTRIC-RESISTANCE WELDED
31	3730126	CA	1972	LONGITUDINAL WELD	MATERIAL FAILURE	ELECTRIC-RESISTANCE WELDED
32	3730127	CA	1972	LONGITUDINAL WELD	MATERIAL FAILURE	ELECTRIC-RESISTANCE WELDED
33	3730148	OK	1973	LONGITUDINAL WELD	MATERIAL FAILURE	ELECTRIC-RESISTANCE WELDED
34	3730163	LA	1973	LONGITUDINAL WELD	MATERIAL FAILURE	ELECTRIC-RESISTANCE WELDED
35	3730164	LA	1973	LONGITUDINAL WELD	MATERIAL FAILURE	ELECTRIC-RESISTANCE WELDED
36	3740015	OK	1974	LONGITUDINAL WELD	MATERIAL FAILURE	ELECTRIC-RESISTANCE WELDED
37	3740019	OK	1974	LONGITUDINAL WELD	MATERIAL FAILURE	ELECTRIC-RESISTANCE WELDED
38	3750076	MI	1975	LONGITUDINAL WELD	MATERIAL FAILURE	ELECTRIC-RESISTANCE WELDED

Lone Star Steel

	Incident ID	State	Year	Leak Origin	Cause of Leak	Type of Steel
1	2830503	TX	1983	LONGITUDINAL WELD	MATERIAL FAILURE	ELECTRIC-RESISTANCE WELDED
2	2830505	TX	1983	LONGITUDINAL WELD	MATERIAL FAILURE	ELECTRIC-RESISTANCE WELDED
3	2830509	TX	1983	LONGITUDINAL WELD	MATERIAL FAILURE	ELECTRIC-RESISTANCE WELDED
4	2830510	TX	1983	LONGITUDINAL WELD	MATERIAL FAILURE	ELECTRIC-RESISTANCE WELDED
5	2830513	TX	1983	LONGITUDINAL WELD	MATERIAL FAILURE	ELECTRIC-RESISTANCE WELDED
6	2830865	TX	1983	LONGITUDINAL WELD	MATERIAL FAILURE	ELECTRIC-RESISTANCE WELDED
7	2830866	TX	1983	LONGITUDINAL WELD	MATERIAL FAILURE	ELECTRIC-RESISTANCE WELDED
8	2831114	TX	1983	LONGITUDINAL WELD	MATERIAL FAILURE	ELECTRIC-RESISTANCE WELDED
9	2831115	TX	1983	LONGITUDINAL WELD	MATERIAL FAILURE	ELECTRIC-RESISTANCE WELDED
10	2831116	TX	1983	LONGITUDINAL WELD	MATERIAL FAILURE	ELECTRIC-RESISTANCE WELDED
11	3700253	TX	1970	LONGITUDINAL WELD	MATERIAL FAILURE	ELECTRIC-RESISTANCE WELDED
12	3700308	NM	1970	LONGITUDINAL WELD	MATERIAL FAILURE	ELECTRIC-RESISTANCE WELDED
13	3700581	OK	1970	LONGITUDINAL WELD	MATERIAL FAILURE	ELECTRIC-RESISTANCE WELDED
14	2710259	LA	1971	LONGITUDINAL WELD	MATERIAL FAILURE	ELECTRIC-RESISTANCE WELDED
15	3720195	CO	1972	LONGITUDINAL WELD	MATERIAL FAILURE	ELECTRIC-RESISTANCE WELDED
16	3730109	KS	1973	LONGITUDINAL WELD	MATERIAL FAILURE	ELECTRIC-RESISTANCE WELDED
17	3730110	KS	1973	LONGITUDINAL WELD	MATERIAL FAILURE	ELECTRIC-RESISTANCE WELDED
18	3730111	KS	1973	LONGITUDINAL WELD	MATERIAL FAILURE	ELECTRIC-RESISTANCE WELDED
19	3730112	KS	1973	LONGITUDINAL WELD	MATERIAL FAILURE	ELECTRIC-RESISTANCE WELDED
20	3730113	KS	1973	LONGITUDINAL WELD	MATERIAL FAILURE	ELECTRIC-RESISTANCE WELDED
21	3730114	KS	1973	LONGITUDINAL WELD	MATERIAL FAILURE	ELECTRIC-RESISTANCE WELDED
22	3730116	KS	1973	LONGITUDINAL WELD	MATERIAL FAILURE	ELECTRIC-RESISTANCE WELDED
23	3730117	KS	1973	LONGITUDINAL WELD	MATERIAL FAILURE	ELECTRIC-RESISTANCE WELDED
24	3750029	OK	1975	LONGITUDINAL WELD	MATERIAL FAILURE	ELECTRIC-RESISTANCE WELDED
25	3760017	TX	1976	LONGITUDINAL WELD	MATERIAL FAILURE	ELECTRIC-RESISTANCE WELDED
26	2770247	TX	1977	LONGITUDINAL WELD	CONSTRUCTION DEFECT	ELECTRIC-RESISTANCE WELDED
27	3780002	TX	1977	LONGITUDINAL WELD	MATERIAL FAILURE	ELECTRIC-RESISTANCE WELDED
28	3740154	OK	1974	LONGITUDINAL WELD	MATERIAL FAILURE	ELECTRIC-RESISTANCE WELDED
29	3770068	LA	1977	LONGITUDINAL WELD	MATERIAL FAILURE	ELECTRIC-RESISTANCE WELDED
30	3770072	LA	1977	LONGITUDINAL WELD	MATERIAL FAILURE	ELECTRIC-RESISTANCE WELDED
31	3770076	LA	1977	LONGITUDINAL WELD	MATERIAL FAILURE	ELECTRIC-RESISTANCE WELDED
32	2770308	LA	1977	LONGITUDINAL WELD	MATERIAL FAILURE	ELECTRIC-RESISTANCE WELDED
33	2821141	LA	1982	LONGITUDINAL WELD	MATERIAL FAILURE	ELECTRIC-RESISTANCE WELDED
34	2821142	LA	1982	LONGITUDINAL WELD	MATERIAL FAILURE	ELECTRIC-RESISTANCE WELDED
35	2821143	LA	1982	LONGITUDINAL WELD	MATERIAL FAILURE	ELECTRIC-RESISTANCE WELDED

Data source: Office of Pipeline Safety web site: http://ops.dot.gov/IA98.htm

Senator GORTON. Thank you, Mr. Tanner.
Deputy Mayor Marshall?

STATEMENT OF CONNIE MARSHALL, DEPUTY MAYOR, BELLEVUE, WASHINGTON

Deputy Mayor MARSHALL. Good afternoon, Senators. Thank you for coming to Washington State to hear about this critically important public safety issue.

My name is Connie Marshall. I am the Deputy Mayor of the City of Bellevue. First, I would like to say how terribly sorry I am to the families here who have lost their children.

When the Olympic pipeline exploded in Bellingham, I could honestly say that I had no idea that this same pipeline ran 11 miles through our city. I can also tell you that since that time I and the rest of my council and our city staff have spent hundreds of hours and thousands of dollars getting up to speed on this issue.

Our Mayor, Chuck Moser, was appointed by the Governor to serve with Bellingham Mayor Mark Asmundson to the Washington State Fuel Accident Prevention and Response Team formed in response to the Bellingham accident.

What we have learned in these past months has not reduced our fears that an accident like the Bellingham tragedy could happen in Bellevue. To the contrary, we have no confidence that the Olympic pipeline is safely operated and maintained within the city. We are greatly alarmed at the complete lack of any automatic shut off valves on this pipeline within the city. From what we have learned, we have even less confidence that the Federal Office of Pipeline Safety is able to provide adequate or meaningful oversight of pipeline operations here or elsewhere in the country.

Our citizens are scared and demanding action, but our ability to respond is extremely limited due to Federal preemption in this area.

In Bellevue, the Olympic pipeline runs through several single family neighborhoods, under a city park, by a middle school and a community swimming pool. There are 11 schools and one community college within one-half mile of the Olympic pipeline in Bellevue. The pipeline also transverses numerous creeks within Bellevue which are the subject of the Endangered Species Act recovery actions. The pipeline also runs under two major freeways, State Route 520, which is the main artery to the Microsoft campus and other businesses in Bellevue and Redmond, and I–90, our major east-west freeway across the state. In fact, the Olympic pipeline runs underneath I–90 less than one-third of a mile from what is one of the busiest freeway interchanges in the state: where I–90 crosses under 405.

Why am I talking about freeways? For a very important reason. Our consulting engineer with extensive expertise in pipelines tells us that the locations where Olympic pipeline crosses under State Route 520 and I–90 are, in fact, the two most vulnerable points of the pipeline within Bellevue. This is because they are the lowest topographical points where gravity exerts the most pressure on the pipe.

One can only imagine the damage to our city and to this region if the pipeline were to fail at these points. Increasing this risk, there are no automatic shut-off valves anywhere within the city. There are only two manual shut-off valves on the 20-inch pipeline at locations within the city. Our engineer tells us that in the event of a rupture, before it could be shut off, the pipeline would drain as much as a million gallons of fuel, four times more than what was involved in the Bellingham accident, and due to our hilly terrain we are also told that the existing technology cannot detect slow leaks in the pipeline. Yet this is within current Federal standards. How can this be allowed within a densely populated urban area?

Olympic last tested this pipeline in 1996 and 1997. They found over two dozen anomalies on the pipeline within Bellevue. Only six of these flaws were repaired. Olympic Pipe Line Company tells us that the other flaws don't meet the Federal threshold for action. Some of these involved as much as 48 percent corrosion on the pipe. Again, how can this be true? How can we possibly explain this to our citizens? How can anyone regard this as an adequate level of protection for our city residents?

While we regard the Olympic Pipe Line Company safety plan as an important step in the right direction, our engineer tells us that the test they proposed will not provide us adequate information regarding the safety or condition of the pipe. He is recommending that we insist on hydrostatic testing unless Olympic can present us with an equivalent alternative. We will be doing so together with asking for better leak detection devices and automatic shut-off valves.

We expect questions will be raised about our ability to make these requests under current Federal law. I would also note that it took us many months to obtain what we regarded as the most basic information from Olympic about the condition of the pipeline within Bellevue, and even the data that we now have provides an inadequate picture. So we are also concerned about the need to impose stricter reporting requirements and public right to know laws on pipeline operators.

In summary, we believe there is an overwhelming case to be made for enhanced regulations in this area. Stricter standards are necessary particularly in urban areas, and because of the limited staffing of the Office of Pipeline Safety, where willing and able, states should be allowed to either act as a proxy for the Federal Government in overseeing compliance or enact stricter safety standards. We strongly support the two pipeline safety bills currently before Congress.

Please work to see that we can ensure safety for our residents. We need better reporting requirements. We need better safety requirements such as safety valves and leak detection devices. We need better Federal funding for OPS. We need assurances that pipeline companies will be required to work now with first responders such as our police and fire departments to develop emergency response plans. We need your help to ensure that the tragedy in Bellingham and the dozen of similar tragedies around the country are not repeated.

Thank you.

[The prepared statement of Deputy Mayor Marshall follows:]

PREPARED STATEMENT OF CONNIE MARSHALL, DEPUTY MAYOR, BELLEVUE, WASHINGTON

Good afternoon Mr. Chairman and members of the Committee, and thank you for coming to Washington State to hear about this critically important public safety issue. My name is Connie Marshall; I am the Deputy Mayor of the City of Bellevue, Washington. Bellevue is an urban city of 107,000 residents located directly to the East of Seattle, on the shores of Lake Washington. We are a major job center in the Puget Sound region, a major retail center, and home to many beautiful single family and multi-family neighborhoods.

When the Olympic Pipe Line exploded in Bellingham, I can honestly say I had no idea that this same pipeline ran 11 miles through our City. I can also tell you that since that time, I and the rest of my Council and our City staff have spent hundreds of hours and thousands of dollars getting up to speed on this issue. Our Mayor, Chuck Mosher, was appointed by the Governor to serve with Bellingham Mayor Mark Asmundsen on the Washington State Fuel Accident Prevention and Response Team, formed in response to the Bellingham accident.

What we have learned in these past months has not reduced our fears that an accident like the Bellingham tragedy could happen in Bellevue. To the contrary. We have no confidence that the Olympic Pipe Line is safely operated and maintained within the City. We are greatly alarmed at the complete lack of any automatic shut-off valves on this pipeline within the City. From what we have learned, we have even less confidence that the federal Office of Pipeline Safety is able to provide adequate or meaningful oversight of pipeline operations here or elsewhere in the country. Our citizens are scared and demanding action. But our ability to respond is extremely limited due to federal pre-emption in this area.

In Bellevue, the Olympic Pipe Line runs through several single family neighborhoods, under a City Park, past a City golf course, by a middle school and a community swimming pool. There are 11 schools and one community college within ½ a mile of the Olympic Pipe Line in Bellevue. The pipeline also transverses numerous creeks within Bellevue which are the subject of Endangered Species Act recovery actions.

The pipeline also runs under two major freeways: State Route 520—which is the main artery to the Microsoft campus and other businesses in Bellevue and Redmond; and I–90, our major East-West freeway across the state. In fact, the Olympic Pipeline runs underneath I–90 less than ⅓ of a mile from what is one of the busiest freeway interchanges in the state: where I–90 crosses under I–405. Why am I telling you about these freeways? For a very important reason: our consulting engineer with extensive expertise in pipelines tells us that the locations where Olympic Pipe Line crosses under SR 520 and I–90 are in fact the 2 most vulnerable points of the Pipeline within Bellevue. This is because they are the lowest topographical points, where gravity exerts the most pressure on the pipe. One can only imagine the damage to our City and to this region if the Pipeline were to fail at these points. Increasing this risk, there are NO automatic shut-off valves anywhere within the City. There are only two manual shut-off valves on the 20 inch pipeline at locations within the City. Our engineer tells us that in event of a rupture, before it could be shut off, the pipeline would drain as much as a million gallons of fuel, four times more than was involved in the Bellingham accident. And, due to our hilly terrain, we are also told that existing technology cannot detect slow leaks in the pipeline. Yet this is within current federal standards? How can this be allowed within a densely populated urban area?

Olympic last tested this pipeline in 1996 and 1997. They found over two dozen "anomalies" on the pipeline within Bellevue. Only six of these flaws were repaired. Olympic Pipe Line Company tells us that the other flaws don't meet the federal threshold for action. Some of these involved as much as 48% corrosion of the pipe. Again, how can this be true? How can we possibly explain this to our citizens? How can anyone regard this as an adequate level of protection for city residents?

While we regard Olympic's Pipe Line Company's Safety Plan as an important step in the right direction, our engineer tells us that the tests they propose will not provide us adequate information regarding the safety or condition of the pipe. He is recommending that we insist on hydrostatic testing, unless Olympic can present us with an equivalent alternative. We will be doing so, together with asking for better leak detection devices and automatic shut-off valves. We expect questions will be raised about our ability to make these requests under current federal law.

I would also note that it took us many months to obtain what we regarded as the most basic information from Olympic about the condition of the pipeline within Bellevue, and even the data we now have provides an inadequate picture. So we are also concerned about the need to impose stricter reporting requirements and "public right to know" laws on pipeline operators.

In sum, we believe there is an overwhelming case to be made for enhanced federal regulations in this area. Stricter standards are necessary, particularly in urban areas. And, because of the limited staffing of the Office of Pipeline Safety, where willing and able, states should be allowed to either act as a proxy for the federal government in overseeing compliance, or enact stricter safety standards. We strongly support the two pipeline safety bills currently before Congress introduced by members of our Washington delegation: S. 2004 and H.B. 3558.

Please work to see that we can ensure safety for our residents. We need better reporting requirements. We need better safety requirements, such as safety valves and leak detection devices. We need better federal funding for OPS. We need assurances that pipeline companies will be required to work now with first-responders such as our police and fire departments to develop emergency response plans. We need your help to ensure that the tragedy in Bellingham, and the dozens of other similar tragedies around the country, are not repeated.

Thank you.

[The attachments referred to have been retained in the Committee's files.]

Senator GORTON. Thank you, Deputy Mayor Marshall.

Mr. Hoggard?

STATEMENT OF CALVIN HOGGARD, CITY MANAGER, SEATAC, WASHINGTON

Mr. HOGGARD. Senator Murray, Senator Gorton, thank you for being here and your serious concern of the issues of the community in regards to pipeline safety and thank you for the opportunity to testify.

The City of SeaTac has 25,000 residents. About 35,000 people come to work in the city each day. Additionally at any given time there are about 10,000 guests living in the hotel rooms in our city. We have each day 75,000 people who come to and from the airport. Large numbers of these people are at risk because of the unsafe pipeline in our community. I share the concerns that you've heard from the other communities that live along the 400-mile pipeline corridor in western Washington.

The basis of our concern is the absence of adequate requirements at the Federal level to operate the pipeline safely, coupled with preemption by the Federal Government of state and local oversight of pipelines, and we have a pipeline operator that seems to be doing the minimum necessary to get along.

The lack of Federal requirements include the lack of requirements for periodic testing, lack of standards regarding the frequency and type of testing, lack of independent oversight of pipeline operations, testing of inspections and followup on deficiencies, the lack of standards for certification and qualification to operate and maintain pipelines, lack of requirements to work with state and local agencies regarding pipeline design, inspection, testing and followup of tests, the lack of requirements to report on inspection results and followup repairs, and a lack of requirements to work in any city with local emergency response agencies such as the police and fire departments including the reporting of incidents in a timely way.

In SeaTac's case, the absence of the ASME requirements we feel has led to poor management on the part of Olympic Pipe Line Company and given us serious concern about our safety. The pipeline in SeaTac is a 12-inch diameter pipe. It operates at 800 pounds per square inch. It pumps millions of gallons of fuel to the airport, where there's a five million gallon storage tank. It runs steeply uphill from the Sea-Tac Airport, or downhill from the Sea-Tac Airport to the Green River valley, a distance of about four miles in a heavily developed urban area that includes the airport, retail businesses, offices, restaurants, hotels, congested arterial highways, freeways and homes, and there's no shut-off valves in that entire length. We could do nothing to prevent the gravity release of thousands of gallons from the pipe into the areas that I've described, with devastating results.

One of our biggest immediate concerns is incident response. Nothing requires the pipeline operators to provide notification of emergency personnel or to develop effective emergency response plans with all the government, police or fire agencies. Olympic has a track record of delayed or no notification of spills.

Despite our substantial efforts today, SeaTac does not yet have an emergency response protocol worked out with Olympic should an event occur in SeaTac. Though they have stated they have done so in their literature, the Olympic Pipe Line Company has not made substantive contact with law enforcement agencies in King County as a whole to plan and train for emergency responses involving pipeline fuel incidences.

Given the poor state of leak detection, and the current operator practices that are in effect, we are relying on witnesses at the scene to call us in order to have an emergency personnel available at the site of a leak in any timely way.

Olympic investigates incidents on their own, and this results in unacceptable delays in local jurisdictions' efforts to contain, isolate, evacuate or otherwise mitigate the effects of an incident. This places responding police and fire personnel and the surrounding community in unnecessary jeopardy.

Another concern that we have is the absence of warning signage. Although third-party damage is one significant factor leading to pipeline failures, the SeaTac Lateral is not currently well marked or posted anywhere within the city to warn potential excavators of its presence. This is one area where local government through our public works and building permit functions could help the pipeline if the pipeline was required to work with us.

Another area of concern that's been mentioned previously is pipeline maintenance. Pipelines are subject to many sources of damage. Some of the particular concerns in western Washington—in our wet northwest—are cathodic erosion which is rust, stress due to steep terrain which is mentioned in the Bellingham situation here, and also stress due to earth movements or the pipe movement, itself, in our wet soils. All of these conditions are found in the City of SeaTac. However, Olympic believes, incredibly, that with proper maintenance and care, a pipeline will last forever. It seems to us that if this it is the approach that is taken, then strong emphasis must be placed on proper testing, maintenance and care. We do not believe Olympic has maintained the 35-year-old pipeline adequately.

Although not required by any regulations, limited resolution pig testing was conducted 4 years ago. We found it very difficult to obtain information from the Olympic Pipe Line Company about the anomalies, thus found, and we have not had an explanation from

Olympic as to the absence of their followup on the anomalies that are within SeaTac.

We lack any confidence in the judgment of Olympic to properly verify and followup to repair anomalies, given their failure to do so previously. This is one concern with Olympic's Pipe Line Safety Action Plan. There is no independent oversight of their plan, and it's not a long-term plan.

We are also not confident in Olympic's long-term commitment to maintenance and safety given their track record, and that would require better marks such as testing requirements, operator certification, exposure requirements and independent oversight. We believe hydrostatic testing is necessary to adequately insure pipeline safety. Hydrostatic testing is a tried and true method, and if properly applied does not damage the line. Since hydrostatic testing is required under Federal law when a pipeline is first laid, and since Olympic and others argue that a properly maintained line will last forever, why should they have any fear of properly designed hydrostatic tests?

We also support state-level independent oversight with teeth to followup on concerns, and it would go a long way toward eliminating our concerns about the safety of the pipelines in Washington. Our community is alarmed and wants action, but we're overly limited by Federal preemption. We would like to see a delegation of pipeline oversight authority to states, and we support the resolution of the National League of Cities adopted in November 1999 calling for amendments to the Pipeline Safety Act to address these concerns.

Our analysis of the Bellingham and our own situation suggests that the same type of disaster that happened in Bellingham could happen and could have occurred and may occur anywhere in Olympic's system. Federal pipeline safety monitoring is so lax that both in the requirements of the pipeline operators and in administratively that we feel no assurance that the appropriate procedures are in place.

We need your help because without these changes to strengthen the independent oversight of pipeline operators, such things as Olympic's Pipe Line Safety Action plan will not have any long-term benefit.

We support the S. 2004 and H.R. 3558 legislation that's in front of you. We encourage their speedy adoption, and I thank you very much and would be happy to answer any questions.

[The prepared statement of Mr. Hoggard follows:]

PREPARED STATEMENT OF CALVIN HOGGARD, CITY MANAGER, SEATAC, WASHINGTON

Senator Gorton and Members of the Subcommittee, I am Calvin Hoggard, City Manager of the City of SeaTac, Washington. I appear and testify today on behalf of the City of SeaTac. Attending here with me today is Mayor Shirley Thompson. She and other Council members in our City share the concerns I will express today. Thank you for your serious interest in the safety of pipelines in our communities and for the opportunity to testify.

The City of SeaTac is a ten-year old city that surrounds Sea-Tac International Airport, south of Seattle. The City has 25,000 residents. About 35,000 people come to work in the City each day. Additionally, at any given time there are approximately 10,000 guests staying in hotel rooms within the City and we have about 75,000 visitors passing through our city each day. A key economic factor in the City's vitality is the Airport. For this reason among others, SeaTac has not joined with six neighboring jurisdictions in lawsuits fighting the expansion of the Airport but has taken a course to cooperate with this essential transportation facility. I mention this to indicate the generally supportive attitude of the City toward federally regulated transportation facilities.

In the case of Olympic Pipe Line we have a serious problem which we share with other local governments along the 400-mile pipeline corridor in Washington and Oregon. We do not believe the pipeline is reasonably safe in our communities. In the SeaTac area Olympic Pipe Line operates an east to west lateral pipeline or pipeline spur coming from the main pipeline. The lateral to SeaTac runs from Olympic's Renton station to the Sea-Tac Airport. The pipeline flow to SeaTac is intermittent. This on-off flow is an added stress to the line from pressure cycling caused by the change in flow. When not making deliveries to SeaTac this lateral sits full of jet fuel under pressure.

Until the June 10, 1999 pipeline spill and explosion in Bellingham, it is fair to say that few communities or government agencies in Washington were particularly aware of the safety issues surrounding hazardous liquid pipelines. Many of these pipelines—and there are thousands of miles of such pipe nationwide—were installed 40 to 50 years ago, prior to significant environmental regulations. The oil companies correctly emphasize that transporting oil and jet fuel by pipeline is much preferable to the usual alternatives—transportation by highway tanker truck and by barge. In SeaTac, where millions of gallons per year of jet fuel are pumped to the airport, an amount projected to greatly increase, a safe pipeline conveyance is clearly preferable to the large number of trucks which would otherwise be traversing our streets.

The main Olympic pipeline was built in 1965 making it almost 35 years old. The SeaTac segment of the pipeline, a 12″ diameter spur, was built later than most of the rest, in 1971. The 12-inch diameter pipeline, which is constantly under pressure (800 pounds per square inch) with jet fuel, runs from the City of Renton westward across the Green River Valley crossing the Green River. The Green River is a major salmon bearing and navigable river flowing into Elliott Bay in Seattle's waterfront. From the Green River the pipeline continues westward running just under a mile along Strander Boulevard between the heavily congested Southcenter Shopping Mall and Target and other stores. At Southcenter Parkway it turns south and runs about half a mile along the Parkway that is lined with commercial and retail development. It then turns west going through the City of SeaTac about a quarter mile under Interstate 5 up a very steep hill above the City of Tukwila (frequently mentioned in media traffic reports as "the Southcenter Hill" due to its common traffic congestion). It then travels about a mile and one-half along South 170th Street, a residential street lined with homes, a corner grocery and one of our fire stations. Next it turns south along International Boulevard, an arterial that is heavily congested much of the time, for about a mile then turns southwestward into a large 5 million gallon tank farm at the International Airport. From the tank farm multiple smaller high-pressure lines run around the Airport to feed various locations traveling under City streets much of the way. The SeaTac Lateral is not currently well marked or posted within the City to warn potential excavators of its presence. There are very few signs, perhaps 3 or so in the entire City. The City itself only recently learned about the presence of the smaller pipelines as we pushed to obtain more detailed information in the aftermath of the Bellingham accident.

Valve placement and control are big issues—if there is a leak, how far back up the line is the place where the spigot can be turned off, and how quickly? In the hilly terrain of western Washington, how do you stop flow draining downhill without properly placed valves? For example, there is only one valve in the SeaTac spur. It is east of Tukwila. Therefore a pipe burst at the foot of the hill near Southcenter in Tukwila would allow the pipe to drain downhill and out of the rupture with no valve to stop it. Some valves are manual, some computer-controlled. In shutting a valve, one must also shut off the flow coming into the system; otherwise pressure will build up. The pressure in the main pipeline is well over 1,000 pounds per square inch, and ranges up to 800 PSI in the SeaTac Lateral, meaning that any flow problem not handled correctly will quickly become a disaster. But adding more valves can upset the flow dynamics of the entire line, and cannot necessarily be done at each City limits. The addition of valves needs engineering analysis and careful computation. State level oversight seems right to attend to both local and system-wide concerns like these.

From Sea-Tac Airport to the Green River, a distance of almost four miles of heavily populated area, there is no shut off valve of any kind. The first one is at the Green River itself. At that location immediately on each side of the river is one valve. A pipeline rupture anywhere along this entire area would seriously risk loss of life and severe environmental and/or property damage. A pipe rupture on Strander Boulevard, for example, would release under high pressure and gravity pressure all the contents of the pipe draining down the hill from SeaTac into a heavily populated shopping area. As I understand it from our fire officials portions of the fuel would likely vaporize into a cloud when released into normal air pressure. The rest would puddle up or flow on the ground. The vapor would be heavier than air so it would also travel along the ground until encountering an ignition source that would cause it to explode with devastating results. Our fire and police could only get people away and watch as the pipeline emptied if we were lucky enough to have any time to have emergency personnel at the scene to do that.

Given the state of leak detection and current operator practices we may not have emergency personnel available at the site of a leak or rupture in any timely way unless we get lucky. Olympic pipeline monitors fuel pressure at a central station in Renton but even if they detected the rupture, without valves they could also do nothing to prevent the gravity release of thousands of gallons from the pipe into the areas I have described. The leak detection system used at the Renton monitoring station only imprecisely monitors unexpected pressure drops in the 400-mile long line. When an unexpected significant drop in pressure is noticed, the first step taken by Olympic is to determine whether or not the pressure monitors are accurate. Then a person is dispatched to go physically see if the pressure drop has occurred because the line is leaking or ruptured. If it is leaking or ruptured then our City emergency personnel are to be notified.

Such delayed notification results in delayed response, so the City has been working with Olympic since the Bellingham explosion to get earlier, immediate notification at the first hint of a problem because time is so critical, the hazards are so great and we are usually closer to the pipeline than Olympic to respond to check for leaks. The City has not received notification from Olympic of prior instances when there have been leaks. Nothing requires this sort of immediate notification.

One of our biggest concerns right now is incident response. We have not had contact with Olympic on this subject until only recently, at our urging. Since it's the local fire departments that will need to be quickly notified, send their trucks to put out fire and provide medical assistance, and local police that will help evacuate an area if necessary due to a spill, this local dialog is absolutely essential. Moreover, the pipeline's emergency response plan must be not just coordinated with, but approved by, the City Fire Department. At the very minimum, the federal law needs to require this type of coordination. Ideally, the federal level will assign to the local level the determination of what type of incident response planning fits the local area.

Although since the Bellingham explosion we have been pressing Olympic and we are pleased there has now begun to be some dialogue, we do not have a coordinated emergency response plan between the operator and the City's emergency response personnel. Nothing requires Olympic to work out such plans with local jurisdictions.

Though they have stated they have done so in literature, the Olympic Pipe Line Company has not made substantive contact with law enforcement agencies in King County to plan and train for emergency responses involving pipeline fuel incidents. A poll of the King County Sheriff and Chiefs of Police in February 2000 showed that Olympic had not contacted any law enforcement agencies to establish a timeline to do so. Olympic Pipe Line Company does not have acceptable policies and procedures, even today, to contact and coordinate with emergency responders of appropriate jurisdiction(s) in cases of suspected or confirmed leaks.

Olympic investigates incidents on their own which results in unacceptable delays in local jurisdictions efforts to contain, isolate, evacuate and/or otherwise mitigate the effects of incidents. This places responding police and fire personnel, and the surrounding community, in unnecessary jeopardy. Pipeline emergency management, like all emergency management is difficult, because emergencies by nature tend to be dangerous, dynamic, complex and confusing. Most emergency responders use the Incident Command System (ICS) to manage emergencies. Timely notice, accurate information, effective communication, organization, and training are essential elements of effective emergency response plans. Federal law must be changed to insure that pipeline companies are part of established emergency response teams.

In addition to the urban routing of a pipeline designed for a rural setting and with no shutoff valves, the absence of signage to thwart third party damage, the weakness of leak detection and the lack of emergency response coordination, there are other reasons for our concern about pipeline safety.

Pipelines tend to move in the ground, the amount of movement depending on the type of soil, stresses on the pipeline, and whether the area is subject to such events as mudslides or earthquakes. Some communities have reported that the actual pipeline location, when checked by probing, is well outside the swath of land (usually 5–10 feet wide) where it was supposed to be. This can be partly due to movement after construction and partly due to lack of map accuracy based on the lack of requirements for engineer-stamped as-built drawings to be provided to local jurisdictions upon construction (i.e., the pipe was not placed in the exact location contemplated by the pre-design drawings). With GIS technology it seems more accurate pipeline location information could be easily provided if required.

There is no industry standard or even agreement as to an appropriate replacement schedule for old pipe. If one buys a house there are rules of thumb for the usual life of various building materials and components. Olympic and others in the oil industry believe that with proper maintenance and care, a pipeline will last forever. It seems to us that if this approach is taken, and it is being taken by Olympic and the pipeline industry generally, then strong emphasis must be placed on proper maintenance and care. Judging by the frequency of major accidents it appears to us that adequate emphasis on maintenance is not happening in the industry in general or at Olympic.

One reason we feel it is unlikely that pipelines last forever is the "cathodic protection" problem. Cathodic protection provides a slight electric current running to the pipe outer surface which resist the tendency for iron to return, or corrode, to its natural state. But cathodic protection is not perfect. Among other concerns, another metal pipe or structure in the ground can interfere with the cathodic protection intended for the principal pipe.

Other reasons for potential damage to pipelines are strain from earth movements and the strain that can result from being under tremendously high, but varying, operating pressures for years and years, which can fatigue the pipeline. The stress points introduced by elevation variations such as in Washington also increases potential damage.

The actions of "Third Parties" are often a major source of damage. While not the dominant source, third party careless actions are a significant source of pipeline damage. Washington has a "one-call" system with signs near buried utilities encouraging contractors and do-it yourselfers to "call before you dig." There is pipeline participation in this program, but there do not seem to be any mechanisms for ensuring that the signs stay in place. More often it appears that people call in after they have hit something. Further, even if the call is made first, there is no guarantee that the company will respond appropriately. For example, prior to the Bellingham accident, Olympic was advised of digging by a contractor in close proximity to the pipe but may not have taken the necessary precautions to protect the pipe's integrity. Persons seeking permits from the City of SeaTac are informed of the pipeline and the need to avoid it, and to contact Olympic. Our recent road and drainage projects on S. 170th had a representative of Olympic present to assure no damage to the line, as did the relevant sections of the International Boulevard projects. We would like to see requirements surrounding these sorts of activities that will better ensure the pipeline operator and contractor's follow up.

The overall federal pipeline regulatory situation appears to be a "Catch-22" since despite the laxness of the federal requirements, "federal pre-emption" prevents states or local communities from having stronger safety requirements of their own which should be tailored to the area's unique environment.

Safety is of course best achieved through adequate preventive measures such as inspection, testing and replacement of defective line segments. Pipeline companies tend to do more than the federal government requires, because the government requires so little. For instance, while there is currently no requirement for in line testing using a smart pig, many companies (including Olympic) use this technique from time to time. But whether the methods chosen by any given company (e.g., frequency of pigging; type of pig used; response to anomalies identified) meet reasonable and appropriate standards is very much in question.

The federal requirements do not include regular testing or inspection, so problems are often only uncovered on an emergency basis or if a report is made if the pipe is accidentally hit during some unrelated construction project. Additionally, federal procedures do not define what an adequate testing process would be. They do not require more stringent standards for older pipelines despite the older age of many lines.

Testing on a regular basis using appropriate methods is important to assure safety. It's also important that the pipeline companies be encouraged to share the results of that testing with states and local communities to ensure accountability. Pipeline testing and follow-up is a major area of concern because there is no routinely required testing of pipelines and no independent third party monitoring of the follow-up to test results. This lack of third party accountability is our major criticism of Olympic's otherwise positive start with their Pipeline Safety Action Plan. Federal requirements should more strongly provide for this third party oversight. They do not at present in any effective way. We support federal legislation which will allow state level independent oversight of routine testing with teeth to follow up on deficiencies by operators.

Various testing devices are used. "Smart pigs," so named because they make a squealing noise as they are pushed through the pipe by the fluids, measure pipe geometry and pipe wall thickness and can infer the existence of various anomalies. There is no oversight of how pipeline operators use (or don't use) the data from smart pig and other testing. The Olympic Pipe Line was smart pigged in 1996–7 throughout the state and over 250 anomalies were found, but before June 10, 1999, according to a July article in the Seattle Times, Olympic had only fixed a few of these and determined that the remainder were insignificant. (One of the supposedly "insignificant" anomalies was at or very close to the point of the June 10 rupture in Bellingham.)

We have this same situation in SeaTac and throughout the rest of the Olympic Pipe Line system. Anomalies have been found in the limited 1996 voluntary testing showing deterioration but assessed by Olympic to not require excavation to verify or repair. In SeaTac there are at least seven anomalies none of which have been verified by physical inspection and none of which was determined by Olympic to require repair. This information was only recently disclosed to us by Olympic after much lengthy effort by the City. Similar experience in neighboring cities with more complete review to date than we have been able obtain in SeaTac has disclosed serious pipeline deterioration with no follow up by Olympic. We fear the same situation exists in SeaTac. The Pipeline Company has scheduled but not yet held sessions with the cities to explain their actions. We should not be in this situation and it does not appear to be unique to Olympic but an industry wide practice. In fact, I understand that federal standards while not requiring testing, allow up to eighty percent erosion of the thickness of a pipe wall before replacement is required. This should be investigated. Regular effective testing should be required against proper standards with independent oversight of the results and follow-up.

Another form of testing is "hydrostatic." This means that the line is emptied of petroleum products and filled instead with water at deliberately higher pressure. Current Federal regulations call for hydrostatic testing only when a pipeline is newly installed. Bellingham required a new round of hydrotesting before re-opening that section of the line. Both hydrotesting and smart pigging have their advantages and weaknesses. Neither is a substitute for the other. We believe both testing approaches should be used and if properly conducted do no harm to the pipeline.

Pipeline companies describe the difficulties with more frequent hydrostatic testing as follows. Such testing means they must stop shipping product to perform the test (unlike pig testing), and must purchase and then treat and dispose of large volumes of water, as well as fully removing water from the pipe after testing, in order not to contaminate the next petroleum products.

Pipeline companies may also claim that hydrotesting is done at unrealistically high pressures, causing failures when none would occur during normal operation. We do not believe that this is true, as pipelines may fail at normal operating pressure for many reasons. It is also claimed that hydrostatic testing damages the line. That is possible if pressures are too high, but experts have told the City that properly controlled hydrostatic tests are "non-destructive" i.e. they cause no damage to the line. In fact, as evidence of its effectiveness, hydrostatic testing is the only test method that can currently determine certain defects. It is worth noting that before Olympic performed the required hydrostatic tests in Bellingham, they first did several repairs to anomalies on the line that smart pigging had previously identified. Even so, the hydrostatic test demonstrated additional pipe weaknesses when leaks occurred during the testing.

The Federal Office of Pipeline Safety (OPS) which administers the national Pipeline Safety Act, is years and sometimes decades behind in implementing the recommendations of the National Transportation Safety Board (NTSB). OPS can, in many cases, act administratively but has often not done so. It is clearly an advantage that the safety body (the NTSB) is independent and reports directly to Congress, but a disadvantage that its recommendations are not mandatory.

OPS is empowered to pick certain states to which it will hand off its authority and did so with a handful of states, including California, Minnesota, New York and Arizona. For reasons that are not clear, OPS subsequently decided that no more states would be granted this opportunity. States need the right to adopt more stringent safety requirements (that are also tailored to the local environs) than OPS has in place at present.

Leak detection as I touched on earlier is another very important issue. There is no federal requirement for pipeline operators to use leak detection systems, and thus no standards for what would comprise adequate leak detection. At present, leaks are mainly noticed because of a drop in pipeline pressure. But if computers and gauges are not operating, a huge leak (hundreds of thousands of gallons) can go undetected for far too long. Olympic presently relies on pressure monitoring in the Renton control center, and over-flying the line every couple of weeks, to detect leaks. Independent, redundant leak detection systems are vital in highly populated and environmentally sensitive areas.

Another leak detection problem relates to slow, persistent leaks. These are too small to be detected by the pressure gauges. But undetected for weeks, months, or even years, they, too, can contaminate groundwater with hundreds of thousands of gallons of petroleum product. For example, a persistent leak in Renton, Washington in 1986 was undetected for over one year and contaminated an aquifer that remains polluted to this day. The recently detected incident in Delware, where 600,000 gallons leaked over twelve years again demonstrated this problem.

In addition to the above, a review should be undertaken to insure that the pipeline system has proper overpressure protection safety equipment in place. Such equipment should not only prevent excess pipeline pressures, but also reduce unnecessary pressure cycling (i.e., pressure surges) that can significantly "age" a pipeline.

Federal regulations are in place to protect workers and the public in or near facilities such as refineries and chemical plants. These regulations, however, do not protect the public living near pipelines. Pipelines are specifically exempted from such "process safety management" requirements intended to ensure that equipment is designed, maintained, and operated safely. One has to have plans reviewed and a permit issued to add a deck on a house. No such technical review or permit is required to build, modify or operate a pipeline.

Many of the problems associated with pipeline safety could be addressed if pipeline operators were held to a standard to be tested for competency and certified to meet minimum qualifications. This is another area not at all uncommon in other critical industries that should be addressed for pipelines through federal legislation allowing states to do this.

We support the legislation now before Congress: S. 2004 and HR 3558 and encourage you to act now to pass these bills that will help stop the repeated preventable leaks and explosions that cause so much safety concern in our community.

Again, thank you for your attention to these issues of vital concern to us.

Senator GORTON. Mayor Asmundson and Bellingham were very fortunate in discovering an expired franchise and using that as leverage, Mayor Tanner spoke a little bit about the threat to cancel a franchise. How about, and I assume that franchise still has a considerable period of time left in its term, Mayor Tanner?

Mayor TANNER. It's 1996, and it's for 10 years. With good conduct, it's extendable for another 10 years.

Senator GORTON. How about Bellevue and SeaTac? What are your franchise lives?

Deputy Mayor MARSHALL. Ours is expiring in 2004, so we will be looking very carefully at Bellingham's model agreement.

Mr. HOGGARD. The City of SeaTac happens to be lucky because management at Olympic Pipe Line didn't realize that they had let their franchise with us lapse. We as a new city thought we'd inherited a franchise that had been entered into with King County, but it had lapsed in 1995. Another indication, I think, of really the dereliction of management that they would allow that to happen, but we are looking forward to going through the same process that Bellingham went through.

Senator GORTON. You've got a good example, don't you?

Mr. HOGGARD. Yes, we do, and we will use every bit of it.

Senator GORTON. Tell me, you all have in your communities local pipelines that are intrastate in nature. Do you get better information and better cooperation with intrastate pipelines that are regulated by a state utilities and transportation commission than you do with the interstate ones?

Mr. HOGGARD. I can speak to the natural gas pipelines in the city. Yes, we do. We have better ongoing dialog. We see people from those utilities from time to time and have discussions with them about issues, and I think we generally enjoy a better working relationship with them. Olympic Pipe Line largely is an intrastate line. It just barely goes over the border of Washington into Portland.

Senator GORTON. Uh-huh.

Deputy Mayor MARSHALL. For months we asked Olympic Pipe Line for information. Receiving it was difficult at best. Finally, we asked our entire state delegation to breakfast in city hall, and we told them about our problem, and it took a letter signed by our state delegation to get Olympic Pipe Line to honor our requests.

What we got was reams of information. Our staff got stacks of it. We termed it X's and O's. Just like Mayor Tanner has suggested, we had absolutely no way to interpret the data, so that's why we had to employ a consultant at great expense to the City of Bellevue so that we could understand what the information even stated, and even with all that data, I just want to say, we are powerless. We have no regulatory power. We have created a map for you. It's in your packet. We have citizens calling up, e-mailing us and saying, "You know, I have to go to a school meeting or a pipeline meeting. One: I might lose my children entirely. They might die or I might have them going to a different school. What do I do?"

Senator GORTON. Is there in your view a difference in risk and a different form of regulation appropriate for natural gas pipelines from liquid pipelines?

Mayor TANNER. I think there's a difference. We get natural gas pipeline ruptures fairly regularly due to construction activity and so on, and thus far we have not had any—the gas company is completely responsive, just immediate response most of the time, and the, we don't think that the danger is there for the same——

Senator GORTON. And by-in-large they're regulated by the state?

Mayor TANNER. Yes.

Mr. HOGGARD. I might add in discussions with our fire personnel, one of the things they pointed out to me was with fuel like jet fuel or gasoline, you know, while it's under pressure, the first little bit that comes out vaporizes like a spray can. The rest will pool up, and of course, both the vapor and the liquid fuel is heavier than air and flows along the ground, and you've heard about the results of what occurs there, and natural gas, I don't believe, has that same characteristic, and so in many respects I think what we're talking about here is even more dangerous and does call for a higher level of regulation.

Senator GORTON. Senator Murray?

Senator MURRAY. Thank you, Mr. Chairman, and it really strikes me as I'm sitting here listening to four Mayors that because of the accident that occurred here, all of you know much more today than you did a year ago and are much more aware of your franchise agreements and getting information is as difficult as it has been. You've gotten some of it.

It is so important that we pass national tougher standards of inspection and training and certification of our operators, because there are literally hundreds of mayors across this country who have no idea to look for what you've looked for. So I think it's extremely important that we pass national legislation to protect those who don't know to ask the questions that all of you are asking.

One question I'd like to ask you is about the right to know provision that we put into the legislation. As I'm listening to you, you have begged and pleaded and asked for information. It sounds to me like you finally got it after exerting some pressure but it was not very understandable. What can we do to make sure that if we require this information to be made available to neighbors or local governments that it is put into language that is easily understood?

Mayor TANNER. If I might respond to that, one of the things that could be done is to require that it be done in the form of a map similar to the map that we provided to you and similar to the map that Bellevue——

Senator MURRAY. You created this from the information, technical information, they gave you?

Mayor TANNER. Yes.

Senator MURRAY. They did not give you the map?

Mayor TANNER. No, they did not. They gave us, as she said, a stack of material and incomplete material, and after we got it, then we had to spend hours and hours and hours to produce these maps on the anomalies and defects.

Mr. HOGGARD. I think that it would be, the delegation to the state is an appropriate level for that, and in that expertise we would be looking for there, and we have contact with state people all the time on environmental issues and so forth, and I think with that expertise it would be sufficient. As you know, pipes sometimes are not buried where the company thought they were, and they can easily track that with GPS technology and so forth where they should readily be able to provide us information. By the way, SeaTac still does not have information about your anomalies and most of the other cities in western Washington do not.

Senator MURRAY. The right to know provision is extremely important, and the companies need to know on a national level they would be required to provide that information to the communities.

On the franchise agreement, Mr. Tanner, you talked about the franchise agreement you've put together but having trouble enforcing it. What can we do perhaps at the Federal level to help with the enforcement of franchise agreements?

Mayor TANNER. I suppose it would require the delegation of authority to the local level—there's Federal preemption in franchise—and as far as I can determine is unenforceable if it conflicts with Federal law. We've got a tight franchise, but I'm still not sure that it would be enforceable. We'd try, of course.

Senator MURRAY. Comments from any of the rest of the Mayors?

Deputy Mayor MARSHALL. Just first of all, I want to thank you both for recognizing that public safety is the No. 1 responsibility of government. That is what our citizens are begging for. They don't care about all the smart pigs and the hydrostatic testing. When I go to citizen meetings, they first and foremost ask me, "Connie, am I safe? Are my kids safe?" So we need your help in getting stricter Federal standards. If we can't, we have no power to enforce anything.

You know, it's interesting, we have no authority to require testing. We asked our consultant if they knew of another accidents other than Bellingham that occurred in a city similar to ours, and we had a public hearing much like this in Bellevue, and he told us about an accident in Corona, California, and I'm sure you're aware of that, where a train wreck occurred, and it nicked the line. Those are some of the words we hear when we ask for information, nick, dent. You know, what does it mean? So Corona didn't have the ability to detect slow leaks either. The pipeline ruptured, and it destroyed two entire city blocks. Our citizens are scared. They're—like you said, there are other communities where citizens don't know about this. Our citizens are informed, and they're not coming to us and throwing rocks and saying, "Do something." They're coming to us and saying two things. They're saying, "What can we do to help you?" We can, we have tons of engineers in Bellevue as you know because of Boeing. "What can we do to help you interpret the data?" and "How can we help you form the emergency response?" They're not coming to us begging for us to do it all. They're coming to us. They're smart. They're informed, and they want to help us, and so we're, I'm just translating that information, because I feel so completely powerless to tell them, there's not much that I can do, but we can all write Senator Murray and Senator Gorton and support them, and that's what we asked them to do.

So again thank you just so very much.

Mr. HOGGARD. With respect specifically to the franchise agreements, I think it would be helpful to require pipeline companies to enter into franchise agreements where they're operating on public rights-of-way and to require them to honor those agreements and, of course, I think we share the concern of Renton about the enforceability, and it really gives us a need for local enforcement powers. The reason that question comes up is because inevitably because there's so little if any Federal regulation. Anything you say in a franchise agreement is going to go beyond that, and so by having stronger Federal regulations, it will help us in our franchise agreements.

Senator MURRAY. Very good. Thank you very much to all of you.

Senator GORTON. Thanks to all of you for your testimony.

We'll now hear from five representatives from state and Federal Governmental agencies.

Again, I bring the hearing to order once again. Ms. Marilyn Showalter, the Chairwoman of the Washington State Utilities and Transportation Commission.

You've gotten pretty good marks so far here today.

Ms. SHOWALTER. Thank you.

Senator GORTON. I'd be happy to hear from you.

STATEMENT OF MARILYN SHOWALTER, CHAIRWOMAN, WASHINGTON UTILITIES AND TRANSPORTATION COMMISSION, OLYMPIA, WASHINGTON

Ms. SHOWALTER. Thank you, Senator Gorton and Senator Murray.

I was appointed to the Washington Utilities and Transportation Commission as chair just a little over a year ago, so this Bellingham explosion came very soon after my appointment, and I will say that we have made pipeline safety one of our highest priorities since then.

I'm a former deputy prosecuting attorney in the criminal division of the King County prosecutor's office. In fact, that's where I met the Governor, Governor Locke. We were prosecutors together, and as a former prosecutor, I firmly believe that laws on the books mean little if they are not enforced. Even the strongest rules are not effective unless our enforcement of them is thorough and consistent.

I have provided the Committee with a detailed description of our safety program, but let me just highlight a few of the facts.

You have, I hope, a yellow sheet. I made it yellow so that you could see it. Do they have that? I guess it's coming to you right now. It's actually a duplicate of what I've already put in the, what you can see there, since I think this can be confusing, is that the utilities and transportation commission regulates intrastate natural gas and hazardous liquids, and that is about in the case of gas, there are approximately 17,000 miles of main lines, those are, generally speaking, the smaller lines—and 246 miles of transmission lines. We do not, as you know regulate interstate natural gas which is about 1,700 miles. And then in hazardous liquids we regulate about 83 miles of hazardous liquid pipelines. We do not regulate 777 miles of interstate pipelines, but as you've heard, there is some interest on the part of OPS in delegating the interstate inspection authority to the state, so that if we had that authority, you can see from there that we would be helping the Federal Government inspect approximately, as I say, 1,700 miles of natural gas lines and 777 hazardous liquid interstate lines.

Our program consists of six full-time safety engineers, and that is a 50 percent increase in the last year. We increased the number of engineers by two since I was appointed. We have about 80 combined years of pipeline safety experience. We spend about 500 days in the field inspecting pipelines and providing technical assistance to the operators.

The program for pipeline safety costs about $700,000 a year. OPS pays 44 percent of this. That is the maximum available funding that any state can receive.

In addition to inspecting pipelines, we also set safety standards for intrastate pipelines. In the case of gas, these standards are more stringent than the Federal guidelines, and these are in the areas of operations, maintenance, construction and reporting.

In the case of hazardous liquids we only received our authority for intrastate and also from our state legislature in 1998. So to get off the ground quickly, we adopted the Federal standards at that point, but we are intending to do a rulemaking and look at whether we should have more stringent standards for intrastate hazardous pipelines.

In the past 9 months, we have done a number of things to improve pipeline safety. By the end of this month we expect to have completed with the Office of Pipeline Safety a comprehensive joint review and inspection of all interstate pipelines. This involves physical tests of leak detection systems, correction control, and other safety factors.

One of the things this joint effort has given our state is some firsthand knowledge by our state engineers of the interstate facilities located within Washington. Also, I was appointed by the secretary of transportation to the technical hazardous liquid pipeline safety standards committee, which is an advisory committee to

OPS, and in addition, one of our members served on the Governor's task force this past 9 months.

As far as the future goes, we may get interstate authority, which would be a major undertaking. We, as I said, are reviewing both our gas and hazardous liquid intrastate rules to see if they should be made more stringent. We have done and will do a number of things to increase third party damage prevention. As you have probably heard, by far the largest cause of pipeline leaks and failures is third-party damage. In our state, for the years 1992 through 1998, third-party damage caused more pipeline leaks and failures than all other causes combined. We will be developing educational and promotional materials aimed at preventing third party damage. The new legislation that our legislature just passed directs us to establish a single one-call number. We now have six in the state, and the legislation also increases penalties for third party damage and failure to report. Finally, we are also seeking the legislature——

Senator GORTON. Does it require an advance call for a location or does it simply just require a report after an accident has happened?

Ms. SHOWALTER. The bill would require, it increases the reporting requirements before you dig, and it also increases the reporting requirements if you have an accident.

We are, the legislature also is setting up a procedure whereby we can do mapping of pipelines throughout the state, but it is a fairly costly proposition, and that will take some time to complete.

Finally as you know, we have worked with the congressional delegation to seek increased authority from the Federal Government and increased standards, a bit of success of which we heard today.

Thank you for the opportunity to testify and to underscore our commitment to public safety.

[The prepared statement of Ms. Showalter follows:]

PREPARED STATEMENT OF MARILYN SHOWALTER, CHAIRWOMAN, WASHINGTON UTILITIES AND TRANSPORTATION COMMISSION, OLYMPIA, WASHINGTON

Thank you Senator Gorton and distinguished members of the Committee for inviting me to testify on the vital issue of pipeline safety. The tragedy in Bellingham has served to focus attention on the need to ensure that pipeline safety laws and practices are the most effective they can be.

Since I was appointed to the Commission just over one-year ago, I have made pipeline safety one of our highest priorities. As a former deputy prosecuting attorney I believe that having strong laws on the books means little if they are not enforced. This certainly is true in the case of pipeline safety. The strongest rules will not be effective unless our enforcement efforts are thorough and consistent.

Today I will provide you with some background on the Commission's pipeline safety program, discuss our efforts over the past year to improve pipeline safety, and speak to several specific initiatives that are forthcoming.

I have provided the Committee with a detailed description of our safety program. Let me highlight some important facts. The Commission is certified by the federal Office of Pipeline Safety (OPS) to adopt safety regulations for and inspect intrastate natural gas and hazardous liquid pipelines. Under that authority, we inspect nearly 17,000 miles of natural gas mains and nearly 250 miles of natural gas transmission pipelines. We do not inspect interstate natural gas transmission lines, which amount to slightly over 1,732 miles in Washington State. For hazardous liquids, we inspect slightly over 83 miles of pipelines but do not have authority to inspect the 777 miles of interstate hazardous liquid pipelines located within Washington State. In both cases, the interstate pipelines are under the jurisdiction of OPS.

Our program consists of six full-time pipeline safety engineers, an increase of two since I came to the Commission. These inspectors have over 80 combined years of pipeline safety experience. Our inspectors spend almost 500 days in the field inspecting pipelines and providing technical assistance to operators. The program costs over $700,000 annually to operate, of which OPS pays 44 percent under the certification program.

In addition to inspecting pipelines, we also set safety standards for intrastate facilities. These state requirements are more stringent than the federal guidelines for natural gas companies. In 1998, our state legislature granted the Commission authority to inspect intrastate hazardous liquid pipelines. We adopted the current federal rules as our initial state standard. During this year, the Commission will review these rules and adopt additional requirements if needed.

During the last year, we have also been active in a number of broad efforts to improve pipeline safety. Foremost among these, by March we expect to complete a comprehensive joint review and inspection of all interstate pipelines in the state with OPS. Our pipeline safety engineers have been in the field with OPS inspectors conducting physical tests of leak detection systems, corrosion control, and other vital safety factors. This joint program has given our pipeline safety engineers detailed first-hand knowledge of the interstate facilities located within Washington.

In addition was appointed by the Secretary of the United States Department of Transportation to the national Office of Pipeline Safety Technical Oversight Committee. Our policy director was a member of the Governor's Fuel Accident Prevention and Response Team. We also have worked closely with the Governor's office, the legislature, and other agencies on pipeline safety legislation.

The aim of all of these efforts is to improve pipeline safety. Our commitment to this goal is substantial and increasing. We will be active agents in seeking to implement the Governor's recommendations for improved pipeline safety, and in ensuring that the laws are enforced.

The review process that followed the Bellingham accident identified a number of steps that can be taken to further improve pipeline safety. Some of these can be accomplished under current authority and funding, others will require legal changes and additional resources.

As I mentioned earlier, we are currently reviewing our pipeline safety rules for both natural gas and hazardous liquids to make sure they are sufficient. As part of this review we plan to require that operational safety plans be submitted to the Commission. This will help ensure that pipeline operators are translating the requirements of the rules into daily operating practice.

We will increase our efforts to prevent third-party damage to pipelines. Third-party damage, primarily through excavation, is by far the largest cause of pipeline leaks and failures. In fact, for the years 1992 through 1998, third-party damage caused more pipeline leaks and failures than all other causes combined. We are working with industry and other government agencies to develop educational and promotional materials aimed at preventing third-party damage. We have also been active in efforts to establish a single, statewide "one-call" system for pipeline locator services, and to increase penalties for third-party damage. We want to make industry and the public more aware to the problem of third-party damage, to make it easier to locate pipelines, and to better enforce requirements to use locator services. Finally, we have been strong advocates for "one-call" centers to adopt minimum operating standards and "best practices," designed to improve consistence and quality.

We are seeking funding for a mapping program that will make it easier for emergency response personnel to respond to pipeline incidents. A funding plan will be completed this year, with the aim of having a mapping system complete by the end of 2005.

Finally, we are working with our congressional delegation to seek changes in federal law to ensure that state and federal pipeline safety efforts support each other and allow more frequent and thorough inspections.

I appreciate the opportunity to testify before you today, and to underscore our commitment to improved pipeline safety.

Natural Gas Distribution Company Leaks Caused by Third-Party Damage

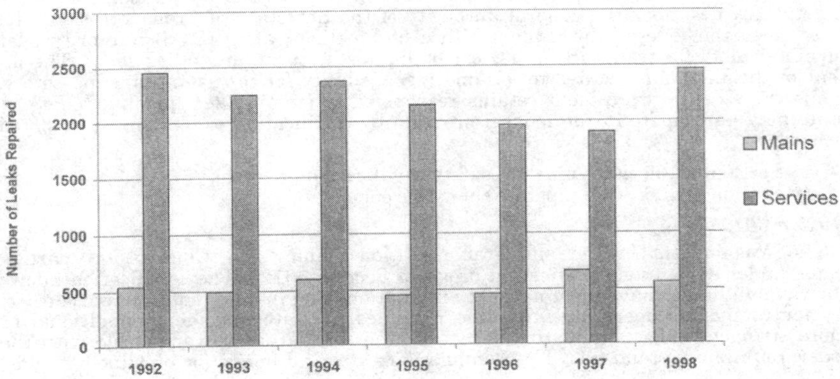

Washington State Pipeline Safety Program

Subject	Gas		Hazardous Liquids	
	INTRAstate	INTERstate	INTRAstate	INTERstate
Miles	17,120 [1]	1,732 [2]	83	777
Delegated Authority from OPS	Yes	No	Yes [3]	No
Standards				
• Meets Federal Requirements	Yes	—	Yes	—
• Stricter than Federal Requirements	Yes	—	No	—
Inspection Authority	Yes	—	Yes	—
Enforcement Authority	Yes	—	Yes	—
Funding for 1999				
• Revenue from Regulatory Fees	$1.2 M	—	$33,950	—
• Federal OPS Grant	.2 M	—	42,800	—
TOTAL	$1.4 M	—	$76,750	—
Staffing	5.5 FTE	—	.5 FTE	—

[1] 16,874 miles/main; 246 miles/transmission.
[2] Transmission miles only.
[3] Hazardous Liquids authority was granted for common carriers in 1996; and private carriers in 1998.

WASHINGTON UTILITIES AND TRANSPORTATION COMMISSION

PIPELINE SAFETY SECTION

OVERVIEW

Regulatory Framework

In 1968, Congress passed the Natural Gas Pipeline Safety Act, now called the Pipeline Safety Law, 49 U.S.C. Section 60101 et seq. Section 60102 of the Pipeline Safety Law describes the general authority of the Secretary of Transportation. The Law gives the federal government Office of Pipeline Safety (OPS) authority over pipeline safety for transporting hazardous liquids, natural, and other gases. The intent of the Law is for states to assume responsibility for *intrastate* pipeline safety, while the federal government retains responsibility for *interstate* pipeline safety. A state may participate in the federal program by certification or agreement as noted below:

- Section 60105 State pipeline safety certification program.
- Section 60106 State pipeline safety agreement.

Section 60105

The Washington Utilities and Transportation Commission (Commission) participates under an annual certification program Section 60105. The certification allows the Commission to have pipeline safety regulations and in-state enforcement authority for companies operating *intrastate* pipelines. A state may adopt additional or more stringent safety standards as long as the standards are compatible with the federal minimum standards. The Commission's state authority for pipeline safety is:

- Chapter 80.28.210 RCW for natural gas, and;
- Chapter 81.88 RCW for hazardous liquids.

The Commission's pipeline safety regulations are found under the Washington Administrative Code (WAC):

- WAC 480–93 Gas Companies, and
- WAC 480–75 Petroleum Pipeline Companies including Hazardous Liquids (carbon dioxide and anhydrous ammonia).

With the certification in place, the Commission's engineers will inspect the following companies:

Intrastate Natural Gas Companies—16,874.0 miles/main, 246.3 miles/transmission

- 4 public service companies
- 3 municipalities
- 1 propane gas distribution company
- 2 propane/air plants
- 7 direct sales
- 250+ master meter/small gas systems

Intrastate Hazardous Liquids Companies—83.4 miles pipelines

- 1 Common Carrier
- 6 Private Carriers

Section 60106

A state may participate under an agreement with the federal agency under *Section 60106* for companies that have *intrastate or interstate* pipelines. The agreement allows the state to act as an *agent* of the federal Office of Pipeline Safety. As agents, state engineers will do the inspections, complete the audits, and file any probable violations with the federal office for enforcement. With an agreement, the federal office directs compliance and enforcement.

The Commission's rules do not apply to interstate companies:

Interstate Natural Gas—1732.1 miles transmission

- 3 Transmission Companies
- 1 Liquefied Natural Gas Facility
- 1 Gas Storage Field

Interstate Hazardous Liquids—777.0 miles

- 4 Trunk lines
- 1 Break Out Tank Company

Pipeline Safety Responsibility

Federal Program—Office of Pipeline Safety

Protecting people and our environment from the risk of pipeline incidents is the responsibility of OPS. This work is shared with state agencies through certification, agreement, or both. OPS administers a range of regulatory protections on several phases of pipeline design, construction, operation and maintenance, research, development, risk management, and damage prevention. The technical Pipeline Safety Standards Committee and the Technical Hazardous Liquid Pipeline Safety Standards Committee are part of the OPS program. The committees provide peer review for carrying out the requirements of the Pipeline Safety Law. The committee may propose safety standards for gas and hazardous liquids to the US Secretary of Transportation. The committee reviews each of the new standards. OPS also requires pipeline companies to plan for spills or leaks and to test their emergency preparedness through planned exercises and other drills. Key to any pipeline safety program is Emergency response by the operator and coordination with local police and fire department. OPS provides training and technical assistance to pipeline companies through the Transportation Safety Institute (TSI) using formal class training and local in-state seminars. TSI also provides education for federal and state inspectors. OPS is implementing a damage prevention study and from that study has identified a list of "best practices" to prevent excavation damage.

Washington Utilities and Transportation Commission's Pipeline Safety Program

The primary mission of the Pipeline Safety Section of the Washington Utilities and Transportation Commission is to inspect pipeline companies and ensure that they are protecting public safety concerning the transportation of natural gas and hazardous liquids. This is accomplished through a commitment to maintaining well-qualified and trained inspectors, field audits, and more stringent rules where required. The Commission currently has six pipeline safety engineers, including two additional full-time engineers authorized in 1999. The inspectors have over 80 years of combined pipeline safety experience.

Each inspector is responsible for field inspections. Typical inspections may include construction, standard, and specialized reviews; operations and maintenance reviews; follow-up audits; responding to incidents; and on-site training for operators.

The Commission has adopted more stringent regulations than the federal minimum safety standards for natural gas companies. These items include:

- filing construction plans,
- proximity and proscribed area reviews for new pipeline operating over 250 psig,
- telephonic and written incident reports,
- increased frequency of leak surveys depending on type of materials,
- installing cathodic protection within 90 days of construction,
- classifying all leaks by grade,
- required leak repair schedules.

The Commission was asked to participate, by the federal Office of Pipeline Safety, in the hazardous liquids program, and staff started the program with authority over one common carrier on January 1, 1996. Legislation for private hazardous liquid carriers was granted and became effective in 1998, and the Commission on January 30, 1999 adopted the WAC, at the minimum federal standard. A Commission priority for the year 2000 includes additional rules for the hazardous liquids program. There have not been any serious incidents or major spills on the *intrastate pipelines* since the Commission started the liquids program in January 1, 1996.

Pipeline Safety Funding

A Pipeline Safety User Fee funds the federal pipeline safety program. Section 60301 of 49 U.S.C. authorizes the assessment and collection of fees. Each operator of regulated interstate and intrastate natural gas *transmission and hazardous liquid* pipeline companies pay a share of the cost of the program based on the number of miles of pipeline. In 1999, the State's appropriation for the combined natural gas and hazardous liquid grant program was $13,000,000 for the base program and $1,000,000 for damage prevention grants. The maximum that a state could receive for 1999 calendar year is 44%. Grant funds are reduced where states do not meet certain jurisdictional and performance requirements and do not have monetary sanctions equal to the federal requirements. Washington's program received the maximum funding level.

Commission Estimated Pipeline Safety Cost

The Commission requested $540,000 for the natural gas and $42,586 for the hazardous liquid program for 1999. The Commission has been authorized to receive the maximum 44% of program cost. The remainder of the Commission safety program is funded by intrastate fees assessed to the utility and common carriers regulated by the Commission.

Pipeline Safety Program—1999 Cost and Grant Allocation

	Total Cost	Grant Allocation	Commission Cost
Natural Gas	$650,127	$238,161	$411,966
Hazardous Liquid	$78,389	$37,564	$40,825

Pipeline Safety Focus

The Commission's Pipeline Safety Section continues to focus on prevention of pipeline leaks. There are five main reasons for natural gas pipelines to leak or fail. They are third-party excavation, corrosion, construction, material defects and outside force. Other causes of failure include cast iron bell-joint leaks and human error. Attached are two charts on Gas Distribution Main and Service leaks, and Third-Party Damage to Mains and Services. The leading cause of pipeline failures is excavation damage, causing 58% of all leaks in Washington State in 1998. When comparing mains to service lines, the damage to service lines represents 82% of third-party damage and mains at 18%. Construction equipment contacts with pipeline can create pipe gouges, dents, scrapes, and cracks. This damage may appear benign at first, but over time, can grow and lead to catastrophic failure. The cost of repair, property damage, and relighting costs has ranged from $234 to over $350,000. The volume of gas or liquid lost or dollar cost for repair may not be an accurate measure of safety. For example:

> A local gas company reported that an excavation contractor was installing a telephone line under a residential driveway. The crew was using an excavation tool called a hole hog that hit and damaged a gas conduit and a ⅝-inch polyethylene service line. The damage caused gas to leak from the pipeline. The excavator did not call the local gas company or the fire department to report the damage or gas leak. A person passing by the site smelled gas odor and reported it. The local gas company and fire department responded on August 12, 1998, and found the leak over the *backfilled* driveway. The gas company removed the backfill and discovered a service line leak that was wrapped with a gum-covered rag placed around the pipe! The contractor's crew had dug up the drive, exposed the leaking gas line, and attempted to stop the leak with a rag. The cost of repair was $800 including overtime charges. The potential for serious injuries and property damage was significant.
>
> On March 5, 1996, in a similar incident, an excavator using the same type of equipment hit a gas line. The escaping gas migrated under the driveway, into the attached garage of a home and ignited, resulting in an explosion and fire. The house and four vehicles were destroyed, with damage estimated to be $350,000. The excavator was digging in a joint trench without first determining the precise location of the gas line. Any uncontrolled leak has potential for serious injuries and large property damage. Prevention is key to pipeline safety.

Washington State Damage Prevention Statute, Chapter 19.122 RCW

Washington State has a damage prevention law found in Chapter 19.122 RCW. The intent of the statute is to assign responsibilities for the location and record keeping of utility location, protection and repair of damage to existing underground facilities, and protection of the public health and safety from interruption in utility services caused by damage to underground utility facilities. The statute provides for civil penalties of $1,000 for each violation and provides for treble costs incurred in repairing or relocating the facility.

The statute requires underground facility owners to join a locator service (One-Call Center). The locator service was set up so that an excavator could call a single number and all underground facility owners would be notified of the planned excavation. The statute requires that excavators call two business days before they plan to excavate, giving the utilities time to respond and surface mark their underground facilities. Six locator services (Call Centers) operate in Washington today. Most states (80%) have a single center and a few states have two centers but no more than three centers. A single-call center could handle all of Washington State's call requests, which are estimated to be about 260,000 per year.

Easements and Rights-of-Way

Permits for land use and rights-of-way for pipelines generally fall under the local jurisdiction of cities or counties and for major projects the Energy Facility Site Evaluation Council. Easements and rights-of-way are not usually considered a pipeline safety regulation and therefore the preemption of the Pipeline Safety Law Section 60104 does not apply. Cities and counties can control where the pipelines are located but are restricted from having pipeline safety regulations. The Commission has natural gas pipeline regulations (WAC 480–93–020 and WAC 480–93–030) that require an intrastate company to file for approval to operate a high-pressure pipeline above 250 psig near people or buildings. These regulations do not and cannot restrict persons from building near a pipeline.

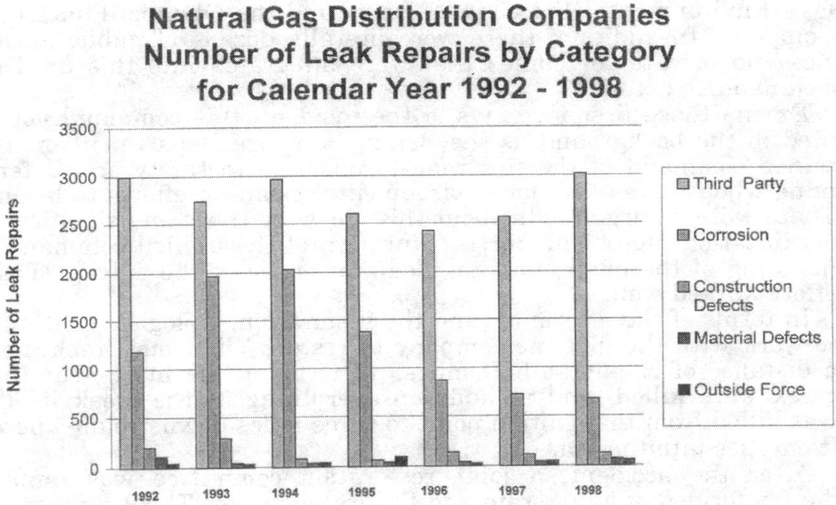

Senator GORTON. Thank you.

Mr. Stohr from the State Department of Ecology.

STATEMENT OF JOE STOHR, SPILL PROGRAM MANAGER, DEPARTMENT OF ECOLOGY, OLYMPIA, WASHINGTON

Mr. STOHR. Thank you Senator Gorton and Senator Murray.

My name is Joe Stohr, and I manage the oil spill prevention, preparedness and response program within the State Department of Ecology, and was also a member of the Governor's task force.

I've been asked to come and give you a brief synopsis of three different areas: First the response to the Olympic oil spill on June 10th, 1999, second to comment on actions that were taken in the aftermath, and then briefly describe to you our regulatory relationship with the Olympic Pipe Line Company.

Let me first begin by clarifying that my agency's response to spills is from an environmental perspective so in the case of the tragic explosion and the fire at Bellingham, our role initially took a back seat, and rightly so, to the fire service and their public safety operations.

Over the years we've responded to a number of large petroleum spills in our state, and our assessment of these emergency responses typically centers around an evaluation of command and control. You can imagine how challenging it is to rapidly mesh together multiple local, state and Federal agencies along with multiple private companies into a cohesive organization in a matter of hours under emergency conditions.

Ecology was very pleased with the speed and selflessness with which the unified command structure was formed and how quickly the emergency operations center was activated in downtown Bellingham. The unified command for environmental response, not to be confused with the fire fighting response, formed within approximately 3 hours of the explosion. It was made up of representatives of the City of Bellingham, the State Department of Ecology, the U.S. Environmental Protection Agency and the Olympic Pipe Line Company. In addition, there were literally dozens of public agencies and private companies that were integrated into this unified command structure.

During those first few days of the incident, this command operated in the background as the Bellingham Fire Department maintained command of the fire zone to ensure site safety and determine when it was safe for environmental clean-up efforts to begin. There were no arguments about this nor were there any significant organizational problems for the duration of the unified command's handling of the environmental clean-up phase of the agency. This effort worked well.

In terms of the aftermath and the restoration, biologists continue to work with the pipeline company to restore Whatcom Creek. For a distance of one-and-a-half miles, all forms of life in or near the creek were killed, and in addition everything in the creek itself, was killed from the ignition point to three miles downstream where it empties into the sea.

After the accident, a joint restoration committee was immediately formed of local, state and Federal agencies. The committee's charge was to identify short-term actions necessary to rehabilitate the stream and allow for the return of salmon and other species.

The Olympic Pipe Line Company was responsive in carrying out these actions which included conducting various studies and monitoring programs, removing residual gas from the sediments, containing further seepage, increasing spawning habitat, hydromulching of sensitive areas to prevent erosion, and developing a draft long-term restoration plan which we should see soon.

In terms of our regulatory relationship with Olympic Pipe Line, in contrast to the team approach immediately following the accident, our regulatory relationship with the company has not always been positive. Again, the Department of Ecology's current authority resides in the enforcement of environmental statutes, and the Olympic Pipe Line Company historically has had a poor performance record in these areas.

Some examples include difficulties in getting the company to submit a quality contingency plan for oil spills, reluctance by the company to discuss spill prevention issues along the main stem of the pipeline, the occurrence of over 50 oil spills of over 825,000 gallons resulting in the assessment of a hundred and fifty thousand dollars in penalties in addition to assessments for natural resources damages, a general lack of attention to spill prevention or response in the recently withdrawn proposal to extend the existing pipeline to eastern Washington, and I guess our perception that the company's corporate culture didn't understand the need for spill prevention and preparedness, and in fact, were far outside industry norms in these areas. So the bottom line here is a lack of change to this culture will probably lead to further significant problems.

On the positive side——

Senator GORTON. Could I stop you there. You said in your view, they're far beyond industry norms. That implies that most of the rest of the industry, other companies have enforced on themselves higher standards than Olympic does in your view.

Mr. STOHR. In this state, that's certainly the case, Senator.

Senator GORTON. OK. Go ahead.

Mr. STOHR. On the positive side, recent changes in Olympic Pipe Line management are welcome, and we see initial progress being made to increase the focus on environmental protection. We hope this trend continues.

That ends my comments and thank you for allowing us to be here.

[The prepared statement of Mr. Stohr follows:]

PREPARED STATEMENT OF JOE STOHR, SPILL PROGRAM MANAGER, DEPARTMENT OF ECOLOGY, OLYMPIA, WASHINGTON

Senator Gorton, members of the subcommittee, my name is Joe Stohr. I manage the Oil Spill Prevention, Preparedness and Response Program for the Washington State Department of Ecology. I have been asked to give you a brief synopsis of the response to the Olympic Oil Pipe Line Company, spill of June 10, 1999, comment on actions taken in the aftermath and briefly describe our regulatory relationship with the Company.

If I may, let me begin by clarifying that my agency responds to spills from an environmental perspective. In the case of the tragic explosion and fire in Bellingham, our role obviously took a back seat to the fire service and their public safety operations and in my characterization of the incident response I won't be speaking to the fire fighting and rescue aspects.

That said, I will add that we have responded to a number of large petroleum spills in our state, and our assessment of these emergency responses typically centers on an evaluation of command and control. It is quite challenging to rapidly mesh multiple local, state and federal agencies, as well as multiple private companies, into a cohesive organization in a matter of hours under emergency conditions. To pull this off, emergency responders nationwide subscribe to a standard organizing principle called the incident command system, and in this case, a variation called unified command.

In short, the department of Ecology was pleased with the speed and selflessness with which the unified command structure was formed, and the emergency operations center activated in downtown Bellingham. The unified command for environmental response (not to be confused with the fire fighting response) formed within approximately three hours of the explosion. It was made up of representatives of the City of Bellingham, the State Department of Ecology, the U.S. Environmental Protection Agency and the Olympic Pipe Line Company. Ultimately, dozens of public agencies and private companies were integrated into this unified command structure.

During the first few days of the incident, this unified command operated in the near background as the Bellingham Fire Department maintained command of the fire zone to ensure site safety and determine when it was safe for environmental clean up efforts to begin. There were no arguments about this, nor were there significant organizational problems for the duration of the unified command's handling of the environmental clean up phase of the emergency.

In Bellingham, we believe the strengths of the environmental response include:

- A collaborative style of decision making in the unified command partnership
- Full participation and integration of local elected officials, minimizing political conflicts

• Successful tapping of statewide and nationwide resources

We believe the lessons learned include:

• The need for incident command training for peripheral agencies and companies.
• The need to smoothly integrate site safety plans into response operations.
• The need to smoothly transition the emergency clean up to the long term restoration of the damage to natural resources.

Aftermath

In the aftermath of this gasoline spill and fire, biologists are working with the pipeline company to restore Whatcom Creek. For a distance of one and one-half miles, all forms of life in or near the creek were killed. Additionally, everything in the creek itself was killed from the ignition point to three miles downstream where it empties into the sea.

After the accident, a Joint Restoration Committee was immediately formed of local, state and federal agencies. The Committee's charge was to identify short-term actions necessary to rehabilitate the stream and allow for the return of salmon and other species. The Olympic Pipe Line Company was very responsive in carrying out these actions which included:

• conducting various studies and monitoring programs;
• removing residual gas from the sediments;
• containing further seepage;
• increasing spawning habitat;
• hydromulching of sensitive areas to prevent erosion; and
• developing a draft long-term restoration plan which we should see soon.

In contrast to the team approach immediately following the accident, our regulatory relationship with the Company has not always been so positive. Again, the Department of Ecology's current authority resides in the enforcement of environmental statutes and the Olympic Pipe Line Company historically has a poor performance record in these areas. Some examples include:

• difficulties in getting the Company to submit a quality contingency plan for oil spills;
• refusal by the Company to discuss spill prevention issues along the mainstem of the pipeline;
• about 50 oil spills of over 825,000 gallons resulting in the assessment of $150,000 in penalty in addition to assessments for natural resource damages;
• a general lack of attention to spill prevention in the recently withdrawn proposal to extend the existing pipeline to Eastern Washington; and
• our perception that the Company's corporate culture didn't understand the need for spill prevention and preparedness and in fact were far outside industry norms in these areas.

On the positive side, recent changes in Olympic Pipe Line management are welcome and we see progress being made to increase the focus on environmental protection. We hope this trend continues.

That concludes my remarks.

Senator GORTON. Mr. Felder, we appreciate you coming all the way across the country to testify for us, and we want very much to hear your testimony.

Your agency has obviously been criticized fairly roundly during the course of the day, and I hope that you will not only give us your formal testimony but make some kind of response to that critique.

STATEMENT OF RICHARD B. FELDER, ASSOCIATE ADMINISTRATOR, OFFICE OF PIPELINE SAFETY, RESEARCH AND SPECIAL PROGRAMS ADMINISTRATION, U.S. DEPARTMENT OF TRANSPORTATION

Mr. FELDER. Thank you, Senator Gorton.

First, I'd like to thank Chairman McCain and you, Senator Gorton, for the invitation to speak to the Committee today.

My name is Rich Felder. I'm the Associate Administrator for pipeline safety in the Research and Special Programs Administration at the U.S. Department of Transportation (DOT).

On behalf of Secretary Rodney Slater, Administrator Kelly Coyner, and the rest of DOT, I'd like to express our condolences to the families of Wade King, Stephen Tsiorvas, and Liam Wood. No community or family should have to suffer the loss that you all have experienced. I hope the steps the department is taking and will be taking to respond to this incident and to prevent others like it may bring this community and these families some comfort.

The experience of this incident has caused all of us to redouble our efforts to prevent such incidents from occurring. In addition, I'd like to thank Senator Murray and Senator Gorton for the extraordinary cooperation and assistance that they and their staff have provided since this tragic event. Both have shown great interest in our program and ways to improve the level of pipeline safety in Bellingham, in the State of Washington, and throughout the nation.

To respond to the tragic incident that took place here last year, the department has worked closely with the State of Washington and affected communities. We understand the need for immediate response and answers following such an incident, and we are working to provide the community with the assurance that they need.

We brought the latest technology to bear in assessing this pipeline. We've worked to restore public confidence, but clearly we have a long way to go. We have required immediate corrective action. We've maintained continuous oversight and have assigned a permanent inspector to Washington State. We are committed to long-term corrective action based on our investigation findings and the findings of the NTSB. We will continue to work with the safety board, with the Department of Justice, the City of Bellingham, and Washington State in taking enforcement action. The line has been shut down, and we will not allow it to reopen until all safety concerns have been satisfied.

Our corrective action plan is comprehensive. We've required reduced pressure on the entire system which provides the same margin of safety as a pressure test. We required pressure testing of portions of the line. We required improvements to valves and the computerized pressure control system. We've required additional training with particular attention to qualification of controllers. We assessed the ability of the pipeline to withstand maximum pressure that could buildup in case of a valve closure. We are requiring internal inspection of the line. It is the testing which we believe will provide the best possible information on any condition that could affect future safety. We will require repair, replacement or hydrostatic testing as appropriate for any such conditions. We want to work with the State of Washington, local communities, and other interested stake holders in every possible way to assure your concerns are addressed.

We're working with the Washington Utilities and Transportation Commission in a comprehensive review of all pipelines in the State of Washington. We are also developing a state-wide map and inventory of all pipeline facilities which will be made publicly available with our report this spring. The information these communities are looking for will be contained in that report. We believe this effort will provide additional assurances for citizens here because we will address all aspects of pipeline safety.

The department has learned a number of lessons from the unfortunate experience in Bellingham that will benefit our long-term plan for the national pipeline safety program. We are committed to continuously improving the pipeline safety program to better address risks to public safety and the environment. Our goal is to prevent incidents like Bellingham from ever happening again. The new regulation we're proposing this month, and it will be out by the 30th of March, will require operators to test their pipelines. That's mandatory testing, and develop a safety plan based on assessing all available safety and damage prevention information——

Senator GORTON. How long will they have to develop that plan?

Mr. FELDER. They'll have a year to develop the plan, and they'll develop it under our watchful eye.

Secretary Slater has set a goal of reducing by 25 percent incidents caused by excavation damage, the leading cause of pipeline failures. We are addressing the human side of the pipeline equation by implementing a comprehensive operator qualification regulation. Building on what we've learned from working with Governor Locke's Fuel Accident and Prevention Response Team and from meetings with city and state officials, in fact, I met with the city council of Bellevue on Friday, the department will help communities better protect pipelines and be informed about the effectiveness of each company's safety programs.

We are pleased that the Washington State Utilities Commission Chair, Marilyn Showalter, has joined our pipeline safety advisory committee. We believe there are opportunities to work better with organizations at the community level that are broadly representative of community needs and capable of making informed decisions about the adequacy of pipeline safety and a community's prevention options.

We're working with SAFE Bellingham, the National League of Cities and other public interest groups to pilot test new approaches to improving communications among communities, operators, and regulators.

Finally, we emphasize our commitment to state partnerships, to providing adequate resources to support state programs and to finding better ways of involving states in promoting activities that enable communities to live safely with pipelines.

These activities include identifying local concerns, investigating those concerns, and identifying ways communities can better protect themselves and the pipelines that traverse them.

In closing, we renew our commitment to this Administration's and Secretary Slater's No. 1 transportation goal, safety, and we pledge continuous improvement in protecting both the public and the environment.

President Clinton's budget for next year seeks an unprecedented level of resources for the pipeline safety program. This increase targets the leading causes of pipeline failures and includes a 50 percent increase in grants to states. With these resources we will work together and produce the level of safety and environmental protection that communities deserve.

Thank you.

[The prepared statement of Mr. Felder follows:]

PREPARED STATEMENT OF RICHARD B. FELDER, ASSOCIATE ADMINISTRATOR, OFFICE OF PIPELINE SAFETY, RESEARCH AND SPECIAL PROGRAMS ADMINISTRATION, U.S. DEPARTMENT OF TRANSPORTATION

I would like to thank Chairman John McCain and Senator Slade Gorton for the invitation to speak to the Committee today. My name is Richard B. Felder and I am the Associate Administrator for the Office of Pipeline Safety (OPS) in the Research and Special Programs Administration (RSPA), U.S. Department of Transportation (DOT). I speak to you today to describe our ongoing efforts to respond to last year's tragic pipeline incident here in Bellingham. In addition, I will describe the pipeline safety program's efforts to keep American communities safe, including our recent work to prevent failures, to enhance environmental protection, to improve data and public access to information, and to respond to emergencies.

On behalf of Secretary Rodney E. Slater, Administrator Kelly S. Coyner and the rest of DOT, I would like to express our condolences to the families of Wade King, Steven Tsorvias, and Liam Wood, and all families around the country that have experienced a similar loss. No community should have to suffer the loss that this community has experienced, and no family should have to suffer the loss that these families have experienced. I hope the steps RSPA is taking to respond to this incident and to prevent others like it may bring this community and these families some comfort. The experience of this incident has caused all of us to redouble our efforts to prevent such incidents from occurring.

In addition, I would like to thank Senator Murray and Senator Gorton for their profound cooperation and assistance that they and their staff have provided since this horrible accident. Both have shown great interest in our program and in ways to improve the level of pipeline safety in Bellingham, Washington, and throughout the nation.

Responding to Bellingham

To respond to the tragic incident that took place here last year, RSPA has worked closely with the State of Washington and the affected communities. We understand the need for an immediate response and answers following such an incident, and we are working to provide the community with the assurance they need.

We have brought the latest technology to bear in the Department's assessment of the Olympic pipeline. While we have worked in the short-term to restore public confidence, RSPA expects our long-term actions to produce significant safety outcomes, and I will address those today. In the short-term, RSPA has required immediate corrective action. We have maintained continuous oversight and have assigned a permanent inspector to Washington State. We are committed to assuring long-term corrective action based on investigation findings. RSPA continues to work closely with the National Transportation Safety Board, the Department of Justice, the City of Bellingham and Washington State. We will continue to take enforcement action as warranted, pending the results of the investigation. The pipeline has been shut down and Administrator Coyner has been clear and firm in her resolve that it will not reopen until all safety concerns are satisfied.

Our corrective action plan is comprehensive. RSPA has required reduced pressure on the entire system. This provides the same safety as pressure testing because it reduces the pressure on the pipeline to the same degree that the pressure test increases the pressure on the pipeline, thereby providing the same safety margin. In addition, RSPA required hydrostatic pressure testing on appropriate portions of the line and improvements to valves and the computerized pressure control system. We have required additional training, with particular attention to the qualification of controllers. RSPA conducted a design review, including assessing the ability of the pipeline to withstand the maximum pressure that could build up in case of valve closure. RSPA required diagnostic tests on the pressure control system.

RSPA required internal inspection of the line. This testing will provide extensive information on the current condition of the line and we believe it will provide the best possible way to detect any conditions which could threaten future safety. RSPA will require repair, replacement, or further hydrotesting as appropriate, for any defects identified.

To conduct the additional testing, the line must be put back in operation during the testing. This will be done at a reduced pressure. Before this occurs, however, RSPA will assess the current condition of the line and the ability of Olympic Pipe Line Company to operate safely. We are close to finalizing our review. After the additional testing is conducted, the line will once again be taken out of service until the Department is satisfied that the line can be safely operated. RSPA will continue to work with the State of Washington, local communities, and other interested stakeholders in every possible way to assure your concerns are addressed.

Comprehensive Statewide Inspection

On October 27, 1999, Secretary Slater directed the Office of Pipeline Safety to work with the Washington Utilities and Transportation Commission (UTC) in conducting a comprehensive review of all pipelines in the State of Washington. To this end, OPS is assessing the safety level of all aspects of pipeline performance, and developing a statewide map and inventory of all pipeline facilities which will be made publicly available. RSPA believes this effort will provide additional safety assurances for citizens here.

Our comprehensive review will address all aspects of pipeline safety, including time of construction, pipe materials, maximum operating pressure, type of commodity transported, internal inspections, failure history, pipe inventory and weld type, maximum flow rate and tank conditions. RSPA also is conducting field verifications of equipment and personnel and visiting pipeline right of ways.

In addition to ensuring the pipeline industry's compliance with all existing regulatory requirements, RSPA is closely reviewing how individual operators address issues of public safety and environmental protection. We will also detail a plan for the continuing safety oversight program for each pipeline system. As I already mentioned, RSPA is also developing a statewide map and inventory of all pipeline facilities which will be made publicly available.

In addition to the information just described, the final report on our comprehensive review will include a description of the public education, liaison, and emergency response planning activities which are expected of every pipeline operator; an overview of the current one-call system, including other aspects of excavation damage prevention programs; and a list of regulatory compliance contacts and executives for each operator. This report will be finalized this Spring, and made available to the public. RSPA believes this comprehensive review effort will provide additional assurances for citizens here, and will serve as a model for pipeline safety activities nationwide.

Long-Term Plan for Pipeline Safety

RSPA has learned a number of lessons from the unfortunate experience in Bellingham that will benefit our long-term plan for the national pipeline safety program. RSPA is committed to continuously improving the pipeline safety program to address risks to public safety and to the environment. Our goal is to prevent incidents like Bellingham from ever happening again.

I would like to take this opportunity to discuss the existing program, and to outline some of our ongoing efforts to enhance the pipeline safety program and provide increased protection of public safety and the environment.

Overall, RSPA has worked to solidify the foundation of pipeline regulation and to revitalize our approach to oversight, both of operators' compliance, and their broader efforts to assure the integrity of the national pipeline system. While our regulations today address the need for pipeline integrity through design, construction, operation, maintenance, operator qualification and response, RSPA will strengthen them further with additional requirements for testing, assessing and addressing the integrity of the national pipeline system. Our current safety standards and oversight practices speak to the need for many forms of inspection and testing, sometimes with very specific schedules. Our initiatives in recent years have focused on further reducing incidents caused by four leading causes of pipeline failure: outside force, corrosion, human error and material defects.

This pipeline safety program has evolved in the 1990's, from an $8 million to a $36 million dollar per year program. RSPA has set its priorities based on the highest risks to public safety and the environment. President Clinton's budget request for next year seeks an unprecedented level of resources for the pipeline safety program, $47.1 million, a 28.5 percent increase above this year's budget. This increase targets the leading cause of failures in all underground utilities, including pipelines, damage associated with excavation. This request includes about a 50 percent increase in grants to states to assist communities with protecting their citizens from pipeline failures by building their damage prevention capabilities and increasing their efforts to oversee the integrity of pipelines. The Administration's budget request includes additional funds for research on outside force damage to locate defects on pipelines at the earliest possible time.

Ongoing Efforts to Program Improvement

RSPA is undertaking a number of efforts, outlined below, to improve the existing program:

Addressing Excavation Damage: First, RSPA will provide strong Federal leadership to address one of the leading causes of pipeline failures—excavation damage. Secretary Slater has set a goal of reducing incidents caused by outside force damage by 25 percent, and RSPA will need everyone to help share in the responsibility for digging safely. RSPA will be providing initial support for a nonprofit organization to continue best practice efforts; to educate the public about how to Dig Safely, using our new national campaign; and to establish a clearing house for damage prevention incident data so we can evaluate program effectiveness. Our national Dig Safely efforts continue, with more than 25 training sessions hosted around the country since June to kick off local campaigns. Many communities are realizing the importance of damage prevention. Also, we must invest in research to better detect and monitor excavation damage. As already mentioned, the Administration's FY01 request includes additional funds for research on outside force damage to locate defects on pipelines at the earliest possible time.

Ensuring Operators are Qualified: Second, we are addressing the human side of the pipeline safety equation. Last year, RSPA finalized a statutory requirement for an operator qualification program to assure a workforce capable of performing safety functions and responding to abnormal conditions. RSPA will work aggressively with operators to review their progress in developing qualification programs. Where progress is inadequate, RSPA will intervene.

Improving Data Availability and Use: Third, a critical lesson RSPA has learned is that we have to improve data collection, and make better use of the information we have. We can do this by assuring integration of information obtained from internal inspections with one call and operating data.

Improving Public Access to Information: Fourth, RSPA is investigating how to help communities better protect pipelines and be informed about the effectiveness of each company's safety programs. We have learned much from the experience of working with Governor Locke's Fuel Accident and Prevention Response Team and from meetings with city and state officials. We are pleased that the Washington State Utilities Commission Chair, Marilyn Showalter, has joined our Pipeline Safety Technical Advisory Committee. While we have requirements today to alert emergency responders about the existence of pipelines, we believe there are opportunities to work better with organizations at the community level that are capable of making informed decisions about the adequacy of pipeline safety and prevention options. We are working with Safe Bellingham, the National League of Cities and other public interest groups on pilot testing some new approaches to improving communications among communities, operators and regulators.

Fostering State-Federal Partnerships: RSPA emphasizes our commitment to State partnerships—to providing adequate resources to support State programs and to finding better ways of involving States in activities that enable communities to live safely with pipelines. These activities include identifying local concerns, investigating those concerns, and identifying ways communities can better protect themselves and the pipelines that traverse them.

The Administration plans to seek additional statutory authority to protect public safety and the environment through an improved pipeline safety program. RSPA looks forward to working with the sponsors of existing pipeline safety legislation in both the House and the Senate, and other Members of Congress, state and local governments, and interested stakeholders on completing a pipeline safety bill this year.

Conclusion

In closing, RSPA renews its commitment to assure continuous improvement in pipeline safety and in protecting both the public and the environment. Thank you, and I would be pleased to answer any questions you have.

Senator GORTON. Before we go on, I'd like you to be explicit. If you get the amount of money recommended in the president's budget, do you feel that you will be able to solve all of the problems that have been outlined here today?

Mr. FELDER. I think that we can certainly address them more successfully than we have in the past.

Senator GORTON. Well, probably do that if we give you a somewhat smaller increase in the budget as well. OK. Thank you.

Mr. FELDER. Can we solve them all? We want to get down to zero the same way everyone in this room does in terms of injuries, fatalities, property damage. Senator, that's what we're working for.

Senator GORTON. Mr. Chipkevich, you have been complimented during the course of the events this day, and we're delighted to have you with us from Washington, D.C. and look forward to your testimony.

STATEMENT OF ROBERT CHIPKEVICH, DIRECTOR, OFFICE OF PIPELINE AND HAZARDOUS MATERIAL SAFETY, NATIONAL TRANSPORTATION SAFETY BOARD

Mr. CHIPKEVICH. Thank you. Good afternoon, Senators Gorton and Murray. I appreciate the opportunity to appear on behalf of the National Transportation Safety Board to report on our investigation into the pipeline rupture and the fatal fire in Bellingham last June.

Accidents are devastating to the victims, their communities, and their families. Most are preventable, but not without a dedicated and persistent effort by industry and regulators to set and enforce high standards.

Attached to my testimony for the record are charts that will help clarify my remarks. The first is a map of the Olympic pipeline system in the northwest. Preliminary data indicate that shortly before the rupture in Bellingham a pump at Woodinville did not start when commanded. A relief valve at Bayview should have worked to relieve upstream overpressure, and failing this, a blocked valve at Bayview should have closed, as we believe it did. Product was pumped into the line at Cherry Point, and the closure of the Bayview blocked valve would have sent a pressure wave back toward Ferndale and Cherry Point. The rupture occurred about midway between Cherry Point and Bayview at the Bellingham water treatment plant. Preliminary data indicates that the rupture occurred at well above the operating pressures, but substantially below the full yield strength for the pipe of this design and size, and even below the maximum allowable surge pressure permitted by regulatory standards. It took several weeks to excavate the ruptured segment, largely because the rupture occurred beneath an area of extensive water piping.

You have also been provided with pictures of the pipe section as initially uncovered. Preliminary inspection of the ruptured segments indicates external damage to the pipe where the failure is believed to have begun, along with additional internal damage. You can see from the pictures that there is some evidence of both external and internal deformities. I would caution you, however, that this information is preliminary, and intensive testing is required before we can be confident of a complete failure. We plan to meticulously test the ruptured pipe segment when possible to determine whether external preexisting damage may have contributed to the rupture, and to understand the consequences of repeated seemingly abnormal closures at the Bayview station blocked valve.

Records indicate that the valve may have closed 50 or more times in the 6-months after it was installed in December 1998. We will examine the reasons for their closures and their possible impact on the upstream pipe's durability. We're also interested in the functioning of a relief valve at the Bayview station. This valve needs to be tested to determine if it is capable of operating within specifications. The design and construction of the Bayview facility also needs close attention to determine if the valve would have been permitted to function correctly in the application.

Given the extensive overlay of water piping, and what appears to be external damage on the ruptured fragment, we are also looking into possible excavation damage. The Safety Board is carefully documenting and analyzing construction work done at the water treatment facility, and Olympic's work to evaluate information developed from internal or smart pig inspections.

We also want to document and analyze the data available to controllers at the time of the accident and to understand their actions during the accident sequence. They seem to have been unaware of the rupture for an extended period of time, and restarting a pipeline after a rupture suggests a significant performance failure. We don't yet know whether this can be traced to training, qualifications, equipment malfunction, poor design in the computer-based control system or some other undetermined reason.

The NTSB wants the answers to all these questions, and we need to know them as soon as feasible. However, necessary tests on the ruptured pipe have not been completed because of a grand jury subpoena requested by the U.S. Attorneys Office and a lack of cooperation from needed witnesses, but ultimately OPS is answerable for the regulatory context in which the pipeline company operates.

The NTSB has for many years argued that periodic verification of pipeline integrity must be a requirement for service. Internal inspections done at the Bellingham pipe identified issues in the field that ultimately failed. That inspection data produced no change, and the regulatory processes did not require a correction.

OPS's response to our 1997 recommendation is in an unacceptable status. Mandatory inspection and testing programs are needed to protect the public and the industry alike. Federal action is long over due.

The same is true of employee qualification standards. The NTSB has had little success convincing OPS that strong training and qualifications are needed for all personnel in safety critical positions.

Mr. Chairman, that concludes my statement. I'd be happy to answer any questions.

[The prepared statement of Mr. Chipkevich follows:]

PREPARED STATEMENT OF ROBERT CHIPKEVICH, DIRECTOR, OFFICE OF PIPELINE AND HAZARDOUS MATERIAL SAFETY, NATIONAL TRANSPORTATION SAFETY BOARD

Good afternoon Senators Gorton and Murray. I appreciate the opportunity to appear before you on behalf of the National Transportation Safety Board to update you on our on-going investigation into the pipeline rupture and subsequent fire that occurred in Bellingham, Washington, last June, and to discuss pipeline safety issues.

As you are aware, the National Transportation Safety Board (NTSB) has been investigating pipeline accidents since 1967, and as a result of those investigations, we have issued over 1,100 safety recommendations that we believe would prevent a recurrence of similar accidents. The steel pipeline that runs through Bellingham is just a small part of the over 160,000 miles of pipelines transporting hazardous liquids nationwide. In 1997, approximately 616.5 billion ton-miles of oil and refined petroleum products were shipped via pipeline, accounting for 64.5% of the oil and refined petroleum products moved throughout the United States annually.

On June 10, 1999, at about 3:30 p.m. Pacific Daylight Time (PDT), a 16-inch diameter pipeline owned by Olympic Pipe Line Company ruptured, and gasoline leaked into the Hanna and Whatcom Creeks in Whatcom Falls Park within the City of Bellingham, Washington. About 5:02 p.m., the gasoline ignited, resulting in a fireball that traveled approximately 1½ miles downstream from the pipeline failure location. Two 10-year-old boys and an 18-year-old young man lost their lives as a result of this tragic accident. Eight additional injuries were documented, along with significant property damage to a single-family residence and the City of Bellingham's water treatment plant. The release of approximately ¼ million gallons of gasoline caused substantial environmental damage to the waterways. Shortly after being notified of the accident, the National Transportation Safety Board launched a team of investigators to the scene. Safety Board personnel were on scene for approximately 5 weeks.

Before providing you with background on the pipeline system and details of the Board's investigation, I would first like to address the considerations the Safety Board must deal with in this investigation. As you may be aware, key pipeline company personnel have refused to respond to Safety Board questions, exercising their Fifth Amendment rights. In addition, the Board has been served with a grand jury subpoena, issued upon application from the United States Attorney's office in Seattle, Washington, which has been extended on several occasions, placing a hold on necessary destructive testing of the sections of pipe which are in the Board's possession. We are working with the United States Attorney's office to resolve these issues.

Before relating the progress of our investigation, let me give you some necessary background on this pipeline system.

Background

Olympic Pipe Line Company is a partnership consisting of Atlantic Richfield Company, Equilon Pipeline LLC (Equilon), and GATX Terminal Corporation, with Equilon acting as the managing partner. Olympic's system extends from refineries in the extreme northwestern corner of Washington State to Portland, Oregon (see Attachment 1). The entire pipeline system is remotely operated from a central control center located in Renton, Washington. From this centralized location, pipeline controllers can monitor key variables, such as pressures and flow rates throughout the entire system. The controllers can also monitor and operate mechanical components, such as pumps and motor-operated valves.

The accident section of pipeline, which was originally installed in 1965, ran from a pumping station near Ferndale, Washington, approximately 37.4 miles southward, to Olympic's Bayview and Allen pumping and storage stations near Allen, Washington. This steel pipeline was constructed of pipe with a wall thickness of 0.312 inches manufactured by Lone Star. In 1966, approximately 725 feet of the pipeline, including the specific section that failed on June 10, 1999, was rerouted to permit construction of a water treatment plant by the City of Bellingham. This new, short section of pipe had the same specified minimum strength and wall thickness as the original, but was manufactured by U.S. Steel.

Attachment 1

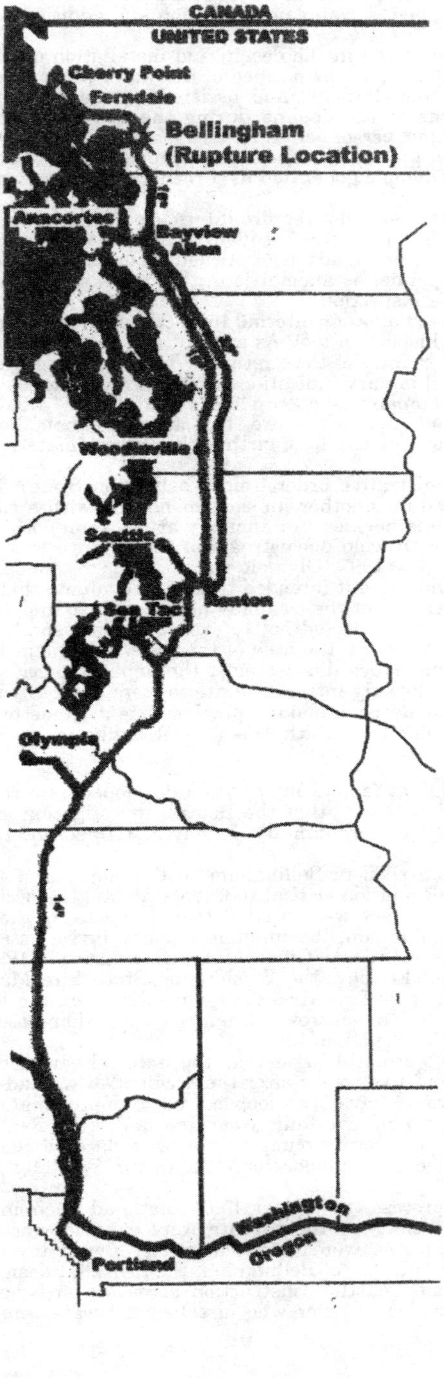

In 1993 and 1994, a contractor working on behalf of the City of Bellingham installed a 72-inch water line across Olympic's pipeline, approximately 20 feet south of the rupture. A new 24-inch diameter water line was also installed and connected to an existing water line 10 feet south of the rupture. In addition, the water treatment plant was being modified, a water pump station and additional smaller crossings were being constructed.

According to personnel involved in the design and installation of the water treatment plant modifications, Olympic was notified of the water plant modifications and associated water pipeline installations, and assisted the design firm with determining the exact elevations of its pipeline during the design phase of the water plant modifications. Olympic personnel were also on site during portions of the water pipeline construction project. Documentation provided to the Safety Board by Olympic includes reports Olympic generated as a result of the water piping installations.

Although Federal regulations do not require internal pipeline inspections, in 1991, Olympic inspected its pipeline from the Ferndale to Allen Stations with a magnetic flux internal inspection tool, or "smart pig." Although anomalies were reported on various segments of the pipeline, no anomalies in the immediate vicinity of the rupture were found during this inspection.

In 1996, Olympic conducted another internal inspection of its pipeline with a similar magnetic flux internal inspection tool. As a result of this inspection, three anomalies were reported in the vicinity of the rupture. While our investigation continues to develop information, preliminary indications are that one of these anomalies, reported by the inspection company as a "possible wrinkle bend," was located in the immediate vicinity of the subsequent rupture. The other two anomalies were located approximately 1.5 feet south of the first girth weld, approximately 10 feet downstream of the rupture.

In 1997, under an administrative order from Washington State's Department of Ecology, Olympic contracted for another internal inspection with a caliper tool specifically to search for pipeline buckles. An anomaly at the same location as the two located south of the first girth weld downstream of the rupture was found as a result of this inspection. In May 1997, Olympic submitted correspondence to the Department of Ecology that indicated it intended to further evaluate this anomaly.

Olympic documents indicate that the company analyzed 1996 and 1997 anomalies mentioned above; however, they elected not to excavate and visually inspect or repair any of the anomalies located in the area of the water treatment plant. Olympic has indicated that these anomalies did not meet the applicable criteria for further action. The Safety Board is looking into what criteria were used. Olympic personnel with direct knowledge of the decision-making process have declined to be questioned by the Safety Board, exercising their Fifth Amendment rights.

NTSB's Investigation

I would now like to highlight factual information developed as a result of our investigation. I would stress, however, that the Board's investigation is ongoing, and that the following information is preliminary. It may be refined and changed as the investigation proceeds.

Upon the Safety Board's arrival in Bellingham on the morning of June 11, 1999, several parties that had information critical to understanding the accident were immediately identified; later, others were invited to participate as the investigation unfolded. Parties to the Bellingham, Washington, pipeline investigation include the U.S. Department of Transportation's Office of Pipeline Safety (OPS), the Washington State Department of Ecology, the Washington State Fire Marshal's Office, the City of Bellingham, Olympic Pipe Line Company (Olympic), the Environmental Protection Agency (EPA), Teledyne-Brown Engineering, Fisher-Rosemount Petroleum, and IMCO General Construction, Inc.

Because the water lines were still in service, the Safety Board did not excavate the ruptured pipe until about two weeks after the accident. If we had excavated immediately, water service would have been jeopardized to approximately 25,000 customers. Sections of the pipe were carefully excavated under the Safety Board's supervision as soon as a new, temporary pump station was placed in service. The segments were then transported to our laboratory facilities in Washington D.C. where they await examination.

During the excavation process, the water lines that had been installed across Olympic's pipeline in the vicinity of the rupture were exposed, and indications of external damage to the pipeline were observed. Safety Board investigators have interviewed personnel from the City of Bellingham, the firm that designed the water plant modifications and managed the construction activities on behalf of the City of Bellingham, as well as the contractor who installed the water piping. However,

Olympic employees who were assigned to inspect the construction activity have also declined to speak with Safety Board investigators.

Safety Board personnel have conducted a thorough visual examination of the ruptured pipeline segment and an adjacent segment that was also removed from the scene. Each of these segments is approximately 10 feet long. Preliminary visual examination of the ruptured pipeline segment has shown that the fracture originated at a gouge mark on the surface of the pipe, and that the gouge at the failure origin was oriented longitudinally along the axis of the pipe. The wall thickness of the pipe at the origin measured between 0.24 and 0.25 inches, a reduction of approximately 20 percent from the original 0.312 inch nominal wall thickness. The overall rupture measured 27 inches longitudinally (see Attachments 2a, 2b, and 2c).* Additional gouge marks and dents were found on the exterior surface of the ruptured pipe segment, and inward protrusions were noted on the inside of the pipe that appeared to correspond to some of the external gouge marks. The external coating on this pipe segment appears to be the original spiral wrap material.

Examination of the second pipe segment noted two dents at the 3:30 and 4:00 positions on the pipe, located 18″ and 22″ respectively, south of a girth weld on this segment. No coating repairs over any of the damage have been noted, and no corrosion damage was observed on the interior of the pipe surfaces or the bare areas of the external pipe surfaces.

Microscopic examination of the fracture face is still necessary to determine whether there are any indications of fatigue near the point of origin. Additional tests are also necessary to determine the microstructure and hardness of the pipe material.

Based upon a review of Olympic's computer pressure data automatically recorded during the accident sequence, our investigators also began to examine the functioning of valves at a newly-constructed pumping and storage facility near Bayview, Washington (the Bayview Products Terminal). Testing was then performed at the request of the Safety Board to determine whether a relief valve at the station had functioned properly. The field testing was inconclusive, and the valve was removed from the pipeline and returned to the Safety Board's facilities for further evaluation.

Preliminary information indicates that pressure began to build within Olympic's Bayview Station as a result of delivery changes underway further down the pipeline system. A relief valve, intended to divert product into a storage tank to reduce the pressure within the facility, had been installed when the station was built in 1998 to protect the piping from overpressurization.

Based upon a preliminary review of pressure information recorded at the Bayview Station, although the relief valve began to function, pressure within the station continued to build, triggering a block valve on the pipeline coming into the station to close. According to information provided by Olympic, when the block valve closed, the pressure on the pipeline upstream of Bayview increased to about 1500 pounds per square inch gauge (psig), and the pipeline ruptured. Information provided by Olympic indicates that the maximum allowable operating pressure was 1370 psig on this pipeline segment. Federal regulations allow pressure surges to 1507 psig. The pipeline theoretically should withstand internal pressure of approximately 2000 psig. The pressure is believed to have reached about 1422 psig at the point of the rupture. After the accident, Olympic recalculated the maximum operating pressure to be 1456 psig at the rupture location.

Preliminary information indicates that the block valve on the pipeline entering Bayview Station had closed over 50 times since the facility began operating on December 16, 1998. On many of these occasions, the valve closure was triggered by a similar pressure buildup within Bayview Station. Our investigators are still evaluating these events to determine the pressures involved and the functioning of the relief valve.

The relief valve was originally ordered with an internal spring set to relieve the pressure at 100 psig. The original Bayview Station design documents called for a set pressure of 740 psig. Olympic subsequently reduced the intended set pressure to 650 psig. In order to modify the valve's set pressure from 100 psig to 650 psig, Olympic ordered a different spring to be installed within the valve's pilot operator. We are looking into, however, whether Olympic replaced a piston and cap as recommended by the valve's manufacturer.

As soon as legal issues have been worked out with the U. S. Attorney's office, the Safety Board will examine the valve to evaluate its performance. Since valves of this type or those with a similar design are commonly used throughout the liquid pipeline industry, it is extremely important fully to understand what occurred.

We also know that the pipeline system design plan for a control valve located upstream of the relief valve intended the valve to be capable of closing completely. The

* The information referred to has been retained in the Committee's files.

valve, however, had an internal stop that prevented it from being capable of stopping the flow of product into the Bayview Station. What effect this might have had on the events that occurred June 10, 1999, is still under investigation.

Shortly after the accident, our investigators also began to evaluate the actions of Olympic's personnel who were operating the pipeline from the Renton, Washington, control center. A preliminary reconstruction of the accident sequence is being performed from a printed summary events recorded within the supervisory control and data acquisition (SCADA) system. A preliminary time line of key events is included at Attachment 3 for your information.

Based on the event logs, we know that flow within the pipeline was restarted at approximately 4:16 p.m., approximately 45 minutes after the rupture occurred. The pipeline was then operated for approximately 17 minutes until the pumps shut down.

Olympic initially reported that a "slowdown" of the computer systems occurred during the accident sequence that affected the ability of the pipeline controllers to change settings on the pipeline system. Olympic further stated that one of its employees may have modified software settings prior to the accident, and may have been working on the computers at the time of the event. A report prepared by Olympic, in response to an OPS corrective action order, acknowledges that the alleged SCADA system slowdown could not be verified or reproduced.

The Safety Board is continuing its analysis of the computer system tapes. Our preliminary review has not identified that a slowdown actually occurred on the day of the accident. Although Olympic has reported to OPS that it has improved its SCADA system by upgrading hardware and balancing workloads between the computer systems since the accident occurred, until we fully understand what happened during the accident sequence, the impact of these changes on future system operations cannot be fully evaluated.

The Board's investigative staff are reviewing substantial documentation provided by Olympic, such as pressure data, design information, construction records, telephone logs and e-mail records, along with the applicable company policies and procedures related to pipeline operations and maintenance, as part of our investigation. However, we will never know what happened within the control center around the time of the accident unless we are able to interview the individuals operating the pipeline when the accident occurred. There are at least four key individuals who may have direct knowledge of the events that occurred in the control room during the accident sequence. Those individuals include two controllers who were on duty at the time of the accident, their supervisor, and a former controller now responsible for maintaining the SCADA system and acting as a relief controller. He was reportedly performing modifications to the computer programming. These individuals are also critical to our investigation into human performance issues, such as training, fatigue, and ergonomics, that may be relevant with this accident. As I mentioned, these individuals and others have declined to talk with us.

The Safety Board is also continuing its analysis of internal inspections conducted by Olympic on the pipeline prior to the rupture, and of the computer operating system and design of the Bayview Station and its effect on the pipeline. We are also hopeful that as the investigative process continues, additional Olympic personnel will be in a position to talk with us.

Safety Issues

Several of the issues being looked into as a result of the Bellingham accident—excavation damage, pipeline integrity, training of personnel—have been concerns of the Safety Board for many years. Excavation damage is the leading cause of pipeline accidents, and the prevention of excavation damage is an issue on the Board's "Most Wanted" list. Recommendations regarding excavation damage were first issued by the Safety Board in 1973, and we are currently investigating several recent accidents in which excavation damage may have played a role.

Our concerns regarding pipeline integrity go back to 1987. As a result of investigations into three pipeline accidents, the Safety Board recommended that the Research and Special Programs Administration (RSPA) require pipeline operators to periodically determine the adequacy of their pipelines to operate by performing inspections or tests capable of identifying corrosion, mechanical damage, or other time-dependent defects that could be detrimental to the safe operation of pipelines. Yet, 12 years after our initial recommendation was issued, there are no regulations that require pipeline operators to perform periodic inspections or tests to locate and assess whether this type of damage exists on other pipelines. OPS has indicated that it intends to enhance enforcement efforts to ensure that pipeline operators who perform internal inspections more aggressively evaluate the results and make appropriate repairs.

ATTACHMENT 3

OLYMPIC PIPE LINE BELLINGHAM, WASHINGTON

PRELIMINARY TIMELINE OF KEY EVENTS

Thursday June 10, 1999

TIME	EVENTS
~3-3:15 p.m.	System Dynamics Occurring due to Delivery Location Changes
3:18 p.m.	Failed Attempt to Start Pump at Woodinville
3:24 p.m.	Oly 2 Computer Operations Down
3:28 p.m.	Oly 1 Computer Operations On-line
3:28 p.m.	Bayview Block Valve Closes
3:28 p.m.	Pumps Shut Down at Cherry Point and Ferndale Stations
3:30 p.m.	Pipeline Ruptures at Bellingham (MP 15.7) and Releases Gasoline
3:34 p.m.	Oly 1 Computer Operations Down
3:48 p.m.	Oly 2 Computer Operations Back On-line
4:16 p.m.	Controller Starts a Pump at Cherry Point and Ferndale Stations
4:24 p.m.	Fire Department Receives its First Call of an Odor from Public
4:29 p.m.	Computer Leak Alert Received at Renton Control Center
4:31 p.m.	Controller Starts Another Pump at Ferndale
4:32 p.m.	Pumps Shut Down at Cherry Point and Ferndale Stations
4:32 p.m.	Controller Initiates Mainline Block Valve Closures
4:34 p.m.	Block Valves at Mileposts 7 and 16 are Closed
5:02 p.m.	Ignition Reported

The Safety Board is concerned that, although the objective is laudable, the efforts may be counterproductive if companies that perform voluntary internal inspections are penalized, thus discouraging them from performing such inspections. It is essential that OPS mandate and enforce a pipeline integrity inspection program for all pipeline operators. The Board's recommendation regarding pipeline integrity was placed in an open-unacceptable status in June 1999.

The need for adequate training of pipeline personnel was also the subject of safety recommendations issued in 1987. The Safety Board recommended that RSPA require operators to develop training programs for pipeline personnel. In October 1998, RSPA published a Notice of Proposed Rulemaking (NPRM) to require pipeline operators to develop a written qualification program for individuals operating pipelines. However, the NPRM did not establish training requirements for personnel, and it allowed companies to evaluate an individual's ability to perform tasks using methods such as oral examinations or observations of job performance. In comments on the rulemaking submitted in January 1999, the Board urged RSPA to amend the rule to include strong training and testing requirements to ensure that employees can properly perform their jobs. We were disappointed that the final rule published in August 1999 was substantially unchanged from the NPRM.

It is unfortunate that some of the issues we have addressed, which have been the subject of Safety Board recommendations, have not been acted on in a timely manner. Each of these issues could be accomplished without legislative action. However, because the Department has not acted, Congressional intervention may be necessary.

Before closing, I would like to take this opportunity to comment on a concern that has been raised by some state officials. As you are aware, state pipeline safety programs are important to help ensure that pipeline system operators comply with minimum safety standards. In fact, state pipeline inspectors who conduct daily inspection activities represent more than 90 percent of the safety inspection workforce. Yet Federal matching funds provided to states have consistently been below the 50 percent level authorized by the Natural Gas Pipeline Safety Act. We have been advised by representatives of several states that funds have not kept pace with demand, and that inadequate funds threaten the infrastructure of the nation's pipeline safety program.

Additionally, we are concerned that while states have many more inspectors than OPS, that OPS is removing states from interstate pipeline inspection programs. State officials have advised that OPS, while previously encouraging states to act as interstate agents, are now having their applications denied. The OPS currently has the ability to utilize these state resources for regular inspection activities through its partnering agreements. It is also critical that comprehensive, consistent, and effective regulatory requirements for interstate pipelines be enacted at the Federal level to protect citizens in all of the states.

For example, in Virginia, approximately 2 million gallons of petroleum products have spilled from pipelines since 1974. In an accident near Reston, Virginia, in 1993, more than 407,000 gallons of diesel fuel spilled from a pipeline into Sugarland Creek, a tributary of the Potomac River. Because of several liquid pipeline accidents that occurred in Virginia, the General Assembly passed legislation in 1994 authorizing the State Corporation Commission to seek interstate agent status from OPS, which would allow state inspectors to inspect interstate pipelines. OPS apparently originally supported this legislation, and for several years encouraged the Commission to pursue interstate agent status. Unfortunately, when the Virginia Commission was ready to accept agent status, OPS denied their application. In fact, states have advised the Safety Board that OPS has effectively halted this program.

We believe state assistance in the interstate pipeline inspection program may go a long way to improving pipeline safety. Because a single pipeline may operate in as many as 10 states, effective Federal oversight is needed to ensure that pipeline operators are meeting minimum safety standards. We believe that Congress needs to closely examine how the states are utilized, funded, and evaluated by OPS. However, for the consistent and effective application of regulatory requirements to interstate pipelines, the authority and responsibility should rest with the OPS.

That completes my testimony, and I will be happy to respond to any questions you may have.

Senator GORTON. Thank you.

Our state fire marshal.

STATEMENT OF MARY CORSO, STATE FIRE MARSHAL, OLYMPIA, WASHINGTON

Ms. CORSO. Thank you, Senator Gorton, Senator Murray. My name is Mary Corso. I'm the Washington State Fire Marshal and the Director of the Fire Protection Bureau with the Washington State Patrol.

I'm pleased with the opportunity to be here to talk about this very important issue relating to pipeline safety, the protection of citizens, the first responder community, law enforcement, fire, EMS, and the environment.

I hail originally from Minnesota where I served 22 years in the fire service, 15 years of which I served as a fire fighter, the remainder, in state service with both the Minnesota State Fire Marshal's office and most recently as the Washington State Fire Marshal.

Twice I have seen the effects of a pipeline incident. First, in 1986 in Minnesota where a pipeline rupture was responsible for two deaths and one serious injury. This rupture took place in a residential neighborhood where people attempting to flee were consumed by an explosion and fire ball leaving no place for escape.

Secondly and most recently in Bellingham where a pipeline rupture allowed hundreds of thousands of gallons of gasoline to flow into a waterway creating a significant tragic event, and three more lives were lost. Three families, an entire community, and the State of Washington bear the grief of this preventable incident.

I am a strong advocate of prevention and preparedness, as I have devoted the last 11 years of my career to protecting life and property through enforcing, engineering, education, and response activities. Protecting people where they live, work and play takes a holistic approach. Protecting people takes passive protection, comprised of safeguards, the majority of which most of us don't see, but exists silently in the background, safeguards such as cathodic protection, monitoring devices and periodic testing. These passive protections need constant vigilance, continuous maintenance, and technically certified operator to insure the integrity of the pipeline.

While you have and will hear significant testimony on these issues, I will focus my remarks on the responses or activities that are also critical in, and vital to protecting our communities.

We are not opposed to pipelines in the State of Washington, and we, in fact, realize that these pipelines may be the safest way for conveyance of liquid and gas fuels. It has been stated numerous times, there are not significant numbers of events involving flammable liquids and natural gas compared to the volume of product that is delivered throughout the state each year. Senator Murray stated it very adequately when she said on average there is one spill per day. This is 365 too many in 1 year.

These numbers of incidents from a response perspective are viewed as low frequency, high risk and have the capability of causing a catastrophic event if a failure occurs. Those who risk their lives in service of their community are in reality community problem solvers. Each time the alarm sounds or the radio crackles, another problem must be solved. This is what we do. There are no second thoughts, no hesitations, just a natural, automatic response to a need in our community.

In the State of Washington each year, first responders—the law enforcement community, fire and EMS—answer over 700,000 calls for help. These include fires, emergency medical incidents, hazardous conditions, law enforcement related activities and a myriad of other emergencies related to public safety. These are the everyday events that we respond to almost automatically, instinctually, if you will.

Our practice and experience comes from these incidents because of their frequency. We concentrate our training, equipment, and planning on these activities as they are what our public expects in terms of protection on a daily basis.

On the other hand, a hazardous material incident due to spills or leaks of flammable liquids and/or the release of natural gas is an infrequent event. Therefore, the necessary training and equipment may not always be current or available. It is very difficult, at best, for many public safety agencies to provide the necessary equipment and training for this very reason. That is why I'm here to talk to you today. Funding, training, and the necessary equipment to respond to these infrequent incidents are vital to our state and our nation's first responders who are called upon to protect life and property and the environment.

The needs of the first responder community are significant. They provide for our ability to maintain the peace, protect the public from fire, and answer the need for emergency medical services, all vital parts of a safe community. Our state's fire service is comprised primarily of volunteers. Up to 71 percent of the emergency responders in our state are volunteers. Of the 650 fire departments in this state, 580 serve populations of 20,000 or less, and they have vast differences in the capabilities within each community.

It was fortunate that Bellingham had a hazardous material response team that was capable of responding to the event that took place on June 10, 1999. Unfortunately, that may not have been the case in many other areas along the pipeline.

The major difficulties facing first sponsors are these. Fire departments are underfunded to deal with hazardous materials. They have significant problems covering the high cost of equipment and maintenance of the necessary equipment for hazardous material response. Greater assistance is needed from the state and Federal level to support local first responder training and equipment needs. The state needs a greater regional response capability for hazardous material teams along the pipeline, with dedicated funding to coordinate training, equipment and supplies.

In Washington State there are only 24 hazardous material response teams associated with fire departments, 12 of which are specialized teams. Yet hundreds of cities are located along the hazardous liquid pipelines in our state. All face serious problems in keeping their staff trained and prepared.

As a member of the Governor's Fuel Accident Prevention and Response Team task force, we identified specific recommendations to assist local first responders. They are: to evaluate the local first responders in communities housing fuel transmission lines for their preparedness; to work in consultation with other state agencies to assess the needs, equipment, and training of local first sponsors; to evaluate the need for training programs to enhance regional incident management teams to assist local responders in managing fuel line pipeline accidents; to amend the Washington State Fire Protection Statute, 48.48 to direct the State Fire Marshal to require that local first responders are immediately notified by pipeline operators if a leak or spill occurs, and; to consult with other agencies to identify the need for and legislative means of achieving consistent application of the National Interagency Incident Response Team.

Pipeline companies need to be active players in this recommendation. It is also important that those who may be called upon at a pipeline incident are able to communicate with each other and operate under a common set of guidelines, terminology and structures.

Additionally, critical to communications is the need for sufficient dedicated radio spectrums identified for public safety agencies to utilize. These systems must be interoperable and provide a system where all responders are able to talk to each other at the scene. That is currently not the case for most major emergencies. This includes first responders, emergency managers, state agencies, and pipeline companies. Standardization, planning, and preparation by all players to prepare for an incident are critical and essential to a safe and positive outcome.

The Governor's Fuel Accident Response Team recommendation and Governor Locke's support for pipeline safety in the State of Washington needs to be emulated at the Federal level to support and protect our public, our nation, and our environment.

In conclusion, I urge passage of the Pipeline Safety Act of 2000, thereby encouraging these same recommendations you have heard today at the Federal level to ensure prevention of future incidents and to ensure our nation's first responders are prepared when needed.

Thank you.

[The prepared statement of Ms. Corso follows:]

PREPARED STATEMENT OF MARY CORSO, STATE FIRE MARSHAL, OLYMPIA, WASHINGTON

Chair Gorton, Senator Murray, and Committee Members:

My name is Mary Corso. I am the Director of the Washington State Patrol Fire Protection Bureau—State Fire Marshal. I am pleased to have the opportunity to address the Senate Commerce, Science, and Transportation Committee regarding this very important issue related to pipeline safety and the protection of citizens, the first responder community (law enforcement, fire and EMS), and the environment.

I hale originally from the state of Minnesota, where I served for 22 years in the fire service, 15 years of which I served as a firefighter; the remainder in state service with both the Minnesota State Fire Marshal's office; and, most recently, as the Washington State Fire Marshal. Twice, I have seen firsthand the effects of a pipeline incident: first, in 1986 in Minnesota where a pipeline incident was responsible for three deaths in a residential neighborhood where people attempting to flee the dangers were consumed by an explosion and a fire ball, leaving no place to escape. Secondly, and most recently, in Bellingham where a pipeline broke, allowing hundreds of thousands of gallons of gasoline to flow into a waterway, creating a significant threat to life safety, property, and the environment. Unfortunately in this tragic event, three more lives were lost. Three families, an entire community, and the state of Washington bear the grief of this preventable incident.

I am a strong advocate of prevention and preparedness, as I have devoted the past 11 years of my career to protecting life and property through enforcement, engineering, education, and response activities. Protecting people where they live, work, and play takes a holistic approach. It takes passive protection, comprised of safeguards, the majority of which we don't see, but exist silently in the background to protect us. Safeguards such as cathodic protection, monitoring devices and periodic testing. These passive protections need constant vigilance, continuous maintenance, and technical operators to ensure the integrity of the pipeline.

While you have, and will hear significant testimony on these issues, I will focus my remarks on the response, or active protection that is so critical and vital to protecting our communities.

We are not opposed to pipelines in the state of Washington. In fact, we realize that these pipelines may be the safest way for conveyance of liquid and gas fuels. While it has been repeated numerous times, there are not significant numbers of events involving flammable liquids and natural gases, compared to the volume of product that is delivered throughout the state each day. However, such operations from a response perspective, are viewed as low frequency, high risk, and have the capability of causing a catastrophic event if a failure occurs.

Those who risk their lives in service of their community are really community problem solvers. Each time the alarm sounds or the radio crackles another problem must be solved. This is what we do. There are no second thoughts, no hesitations, just a natural automatic response to a need in our community.

In the state of Washington each year the first responders (law enforcement, fire and EMS) answer over 700,000 calls for help. These include fires, emergency medical incidents, hazardous conditions; law enforcement related activities and a plethora of other emergencies related to public safety. These are the everyday events that we take care of almost automatically—instinctively, if you will. Our practice and experience comes from these incidents because of their frequency. We concentrate our training, equipment, and planning on these activities, as they are what our public expects in terms of protection every day.

On the other hand, a hazardous materials incident due to spills or leaks of flammable liquids and/or the release of natural gas are an infrequent event. Therefore, the necessary training and equipment may not always be current or available. It is very difficult at best for many public safety agencies to provide the necessary training and equipment for this very reason.

It is for this very reason I am here to talk to you today. Funding, training, and the necessary equipment to respond to these infrequent incidents are vital to our State's and the Nation's first responders who are called upon to protect life, property, and the environment. The needs of the first response community are significant; they provide for our ability to maintain the peace, protect the public from fire, and answer the needs for emergency medical services—all vital parts of a safe community.

To complicate matters, our State's fire service is comprised primarily of volunteers, making up 71% of the emergency response community. Of the 650 fire departments in the state, 580 serve populations of 20,000 or less, with vast differences in the capabilities within each community. It was fortunate that the City of Bellingham had a hazmat response team that was capable of responding to the event that took place on June 10, 1999. Unfortunately, that would not have been the case in many other areas along the pipeline.

The major difficulties facing local first responders are:

- Fire departments are under-funded to deal with hazardous materials; they have significant problems covering the high cost of purchase and maintenance of necessary equipment for a hazardous materials response.
- Greater assistance is needed from the state and federal level to support local first responder training and equipment needs.
- The state needs a greater regional response capacity for hazardous material teams along the pipeline—with dedicated funding—to coordinate training equipment and supplies. In Washington State there are 24 publicly funded hazardous material teams, 12 of which are specialized teams. All face serious problems in keeping their staff trained and prepared.

As a member of Governor Locke's Fuel Accident Prevention and Response Team task force, we identified specific recommendations to assist the local first responders. These recommendations directed the State Fire Marshal to:

- Evaluate preparedness of local first responders in communities housing fuel transmission lines.
- In consultation with the Military Department's Emergency Management Division, the Department of Ecology, and local agencies, the Fire Marshal should conduct a needs assessment of local first responders' readiness and equipment needs particularly relevant to fuel transmission pipelines. This should include consideration of the costs and benefits of meeting identified needs.
- Establish a temporary position to develop training programs for local first responders—police, fire, and emergency medical service staff and volunteers—to deal with pipeline accidents. This person should coordinate with pipeline operators to identify their role in providing the training and to identify the timetable and costs for providing this training to first responders in communities housing transmission pipelines. The program should also address community education and response, including support materials and handouts.
- Evaluate the need for a training program to enhance regional incident management teams to assist local responders in managing fuel pipeline accidents.
- To amend the State Fire Protection Statute (RCW 48.48) to direct the State Fire Marshal to require that local first responders are immediately notified by pipeline operators of any leak or spill, and to;
- Consult with other agencies to identify the need for and legislative means of achieving consistent application of the National Interagency Incident Management System (NIIMS).

It is important that those who may be called upon in a pipeline incident are able to communicate with each other and operate under a common set of guidelines, terminology and structure. Additionally, critical to communications is the need for sufficient dedicated radio spectrums identified for public safety agencies to utilize. These systems must be interoperable and provide a system where all responders are able to talk to each other on the scene. This includes first responders, emergency managers, and the pipeline companies. Standardization, planning and preparation by all players to prepare for an incident are critical and essential to a safe and positive outcome.

The Governor's Fuel Accident Prevention and Response Team recommendations and Governor Locke's support for pipeline safety in the State of Washington needs to be emulated at the federal level to support the protection of our public, our Nation and our environment.

In conclusion, I urge the committee to support the "Pipeline Safety Act of 2000," thereby encouraging these same recommendations you have heard today, at the federal level, to ensure prevention of future incidents and to guarantee that our nation's first responders are prepared and ready when needed.

Thank you.

Senator GORTON. Thank you.

Ms. Showalter?

Ms. SHOWALTER. Yes.

Senator GORTON. What if we were to amend the present law and Senator Murray's bill to say that where an interstate pipeline where more than 90 percent of the interstate pipeline is associated in a single state, that state would have full regulatory authority over the pipeline, would you welcome that delegation of authority? Could you carry it out, and would we have a better result with Olympic?

Ms. SHOWALTER. We would welcome that authority. I think we could not carry it out immediately without, of course, putting more resources into our inspection program, but whether we get the authority through the delegation of OPS of its authority or whether we were given our own authority, we will, in general, have the capability to do it, but we would have to add several more engineers and inspectors.

Senator GORTON. Do you think you could then do a better job?

Ms. SHOWALTER. I think we would do a better job, probably, because we care more about our state, but I have to say I think equally important is I think we would do a better job because we would put more resources into it.

If you look at the number of inspectors that we have or the number of inspectors that we would add if we get interstate delegation, it's probably three or four inspectors, and so we would have a total, say, of, with our intrastate authority of about 10. That is way more than——

Senator GORTON. —than the one?

Ms. SHOWALTER. Than the one, right. So I think both elements are important, what are the resources devoted to it, which could be at the Federal level, but we're willing to devote those resources, and then the other is who has the legal authority.

Senator GORTON. The rest of the panelists will pardon me if all or almost all of the rest of my questions are for Mr. Felder.

Mr. Felder, you've been here all afternoon. You heard Mr. King's question. Given Olympic's record, why is its pipeline continuing to be operating below Bayview?

Mr. FELDER. The reason that we have not shut that pipeline down is because we have already reduced the operating pressure on that portion of the line to 80 percent, which is as I said in my statement, is the equivalent of pressure test. It's the same thing as pressure testing the line, you get the same margin of safety, and we took a look at the safety and the operating history of the line below Bayview, and all of the other information that we have at hand based on our inspections of the Olympic pipeline and deemed it safe to operate, which is not to say that we've given it a final check mark. We certainly have not. We would not allow the pipeline below Bayview to return to a full level of service until we have conducted a full gamut of internal inspections with all of the high resolution tools and the best available technology. So we want to give the citizens of this state what is the equivalent of a brand new pipeline, something that they have confidence in, something that they understand the condition of, and something that they can live with with a much higher level of comfort than I've seen in this room.

Senator GORTON. Then the concerns expressed by the officials from Bellevue, Renton, and SeaTac are without much foundation? You're convinced that their cities are appropriately protected?

Mr. FELDER. Not only have we reviewed the internal inspection runs that they are concerned about, but we've also had outside experts take a look at it, because I know that certainly in this environment our credibility isn't the highest. So we may as well get somebody in that everybody has confidence in. These things have been looked at.

For example, in Renton, we've been in touch with the Mayor there. We've gotten back to him on the evaluation of what he was concerned about. The areas that we had a common concern about in fact had already been addressed, and if in fact after these more sophisticated tools, if they show anything of concern that needs to be examined and then repaired, replaced, whatever is required, we are going to order that. It falls within the gamut of the corrective action that we're going to take, and we intend to take it.

Senator GORTON. On another subject, why has there been no action positive or negative over a period of years on so many of the National Transportation Safety Board's recommendations?

Mr. FELDER. Well, Senator Gorton, I can only say if you take a look at where we are today, we've got a couple of old recommendations. It's absolutely true. There's probably five or six of them that predate 1995 that are still open that are on the books, and all of those recommendations we believe will be satisfied when we put this rule out on the 30th. They have to do with periodic testing. Some of them where not as pointed as this. I mean the oldest one was put out in 1987, and it concerned hydrostatic testing, and its safety goal was to ensure the integrity of pipelines, and when we addressed that recommendation, we tried to address it in a way that was as responsible as possible, which was to take all of the pipelines in America that we thought had susceptibility to the type of defect that a pressure test would expose and order the pressure testing of those lines, and we did that, and we feel that by the time that protocol is completed, and it will be completed within the next year, any pipeline that has that kind of a defect will have been addressed, and I think the good news is that there is now a new, more sophisticated internal inspection device which can actually detect the kind of seam problems that a hydrostatic test is designed to expose, and that tool we are ordering for the first time to be used here in the State of Washington. We've worked on the development of that over the past 3 years.

Senator GORTON. You've listened today, and you've listened before today to a wide range of complaints by local officials here. We and the Congress keep hearing rumors or confusing reports that the Administration of the Office of Pipeline Safety may be seeking to amend the Federal law by revoking the agency authority that it's given to four states and now offered to the State of Washington. Is there anything to that rumor?

Mr. FELDER. Well, I'm very glad you raised that, Senator, because this is a great time to clear the air on that.

The traditional interstate agent policy that we had was one that was developed a number of years ago when the office that I run today had a budget that was in the single digits of millions of dollars. It had a staff that was about a third of what we have today, and it had the same set of responsibilities. So really out of a sense of urgency and out of lack of resources we developed a program to have the states conduct interstate inspections on interstate pipelines on our behalf. We feel it was imperative because otherwise we could never cover the interstate systems with the resources that we had.

Fortunately or unfortunately as you may assess it, after the tragic accident in Edison, New Jersey in 1994, Congress doubled our budget. We went from what had finally grown to 17 million to 37 million dollars in 1 year, and since the budget was raised to thirty-seven, to that 37 million dollars, we took those resources as the Congress provided them to us and hired a large number of new inspectors. We went from 25 inspectors up to 55 to cover the interstate systems, and as we acquired those resources and were able to do those inspections, we actually put a freeze on creation of new state authority to conduct inspections on our behalf.

I will add that states weren't actually beating down the door for this, but there were some states who were interested who we did not authorize, including the State of Texas. However, we have entered into agreements with the State of Texas periodically to do any number of inspections and special construction investigations either apart from us or in conjunction with us. We have always partnered effectively with the state pipeline safety programs, but what we're looking forward to, and let me give you the second part of it, we have a legislative proposal which I hope will reach you by the end of this month, and what we're looking to do, we're not ending any programs where we don't have the ability to conduct those inspections, and if there is a concern that we are not inspecting often enough or thoroughly enough in a state, we're very happy to respond to that, but the program that we have in our proposed legislation actually not only reaches out to states, asks them for their local concerns, both to their population and their environment and has us put a work plan together to address those concerns through inspections and other activities, but it also fully funds that. So what we're looking forward to is support from everyone for the dollars and the resources to partner effectively with states, not just have them conduct inspections using our standards, but to develop work plans based on local and state concerns and conduct those inspections.

Senator GORTON. Well, that's a nice long answer, but I'm not at all sure that it answers the question that I asked.

Just put it very simply——

Mr. FELDER. I can give you a one word answer. Are we ending the program? No.

Senator GORTON. And if we give you the money that the president's budget has asked, you will not back away from these delegations?

Mr. FELDER. Well, I think that we should, we should continue that debate, and I think that that's what the—no, really. I think that's what this reauthorization debate is all about, whether it's a better idea to have each state conduct the pipeline safety investigations or to have a Federal oversight of the pipeline safety investigations with adequate resources or something in-between. I think that's really what this is about, and we want to be involved in that. If there's a determination that we should just be a grant giving agency and have the state perform all the inspections, that's a decision that can be made. If it's something that goes to the other end of it, or if it's somewhere in-between, what we think is important is to really involve states in decisionmaking and partnership in pipeline safety concerns not just to use them as our extra work force.

Senator GORTON. What kind of history do you have? You've got four states that have had this inspection authority for some time. Is the record of incidents with those states better, the same or worse than the other 46 states?

Mr. FELDER. Actually, we took a look at that, and there's no material difference.

Senator GORTON. Senator Murray?

Senator MURRAY. Thank you, Mr. Chairman.

Following up on what Senator Gorton asked you, Mr. Felder, what would be the arguments against allowing states to develop their own inspection regulations?

Mr. FELDER. Well, if we have the responsibility to oversee the safe operation of pipelines, let me give you one, an example of one pipeline, Colonial Pipeline. It runs from Texas to New Jersey, and its control center is in Atlanta, Georgia. I guess the question is who would set the standards, and the way we see it we would rather work with all of the states up and down that pipeline corridor, identify their concerns and then putting together a comprehensive program which we honestly believe we have in the Colonial setting to make sure that that pipeline is operated safely. So we don't, we are not, we don't want to count the states out, but we want them to give us what we need to help protect their citizens. I believe that everyone in this room and in this state knows more about what's important locally both population and environmentally than we do.

Senator MURRAY. So your concern is more for the states where a pipeline goes through two, three, four, five states at a time and who would set the regulations and rules, and that what Senator Gorton suggested a moment ago, if 90 percent of a pipeline operates within one state, would you have concerns about that if the state had authority?

Mr. FELDER. We don't regulate the transmission lines that are in one state.

Senator MURRAY. Right.

Mr. FELDER. So if you wanted——

Senator MURRAY. But the majority is in Washington State, partially in Oregon, would that be of concern to you if Washington State had particular regulations or operated the pipeline?

Mr. FELDER. It might be a concern to Oregon.

Senator MURRAY. Well, let me continue on, Mr. Felder. I think you've heard the charges and the insinuations.

Mr. FELDER. Oh, yes.

Senator MURRAY. But I wanted to ask you specifically because we keep hearing the concern that the Office of Pipeline Safety has more the interests of the industry in its purview in making decisions rather than citizens. How do you respond to that?

Mr. FELDER. I respond to that by saying absolutely not. I respond to that by saying that we are out there every day inspecting, enforcing and trying to get the highest possible standards in place. Some of these things take time. We've been criticized for taking too much time. We'd like to quicken the pace, but in no way do I accept any suggestion that we're either favoring or working on the behalf of the folks that we regulate.

Senator MURRAY. Mr. Chipkevich, you've been critical for a long time of the Office of Pipeline Safety and many of the regulations. You've listened to the testimony today. Do you think progress is being made? Do you think we can feel perhaps Bellingham made a difference and things are going to be better? What's your view from listening to all this?

Mr. CHIPKEVICH. I hope Bellingham does make a difference, but we've seen proposed rules in the past that haven't been effective, or that haven't gone as far as the Safety Board felt the rulemakings needed to go.

An example is in the qualification or training requirements for people operating pipelines. We believe that final rule is insufficient even after many years, and now there's a final rule.

We know there's work being done to improve regulations for corrosion inspection and for the testing of pipelines. But we haven't seen what the final rule is and if it's strong enough. No, we haven't seen the change yet, but I hope this accident does provide the impetus to make those strong changes.

Senator MURRAY. In our legislation that we are proposing at the Federal level, we increase those standards nationally, and I suppose that you would support that?

Mr. CHIPKEVICH. I've looked at the legislation, and I think there's some very strong, important elements in that legislation. I think that some of those standards could have been certainly implemented without legislation, but maybe legislation is required in order to get what we're finally looking for. We can certainly provide you some specific comments for the record on that.

National Transportation Safety Board staff met with your Committee staff to discuss Safety Board recommendations for pipeline safety improvements related to the proposed legislation. A review of our data base reveals that safety recommendations issued to the Research and Special Programs Administration can all be accomplished without legislative action. However, a number of critical safety issues in the proposed legislation have not been acted upon and may warrant Congressional intervention.

Since 1987, the Safety Board has urged that RSPA require pipeline operators to verify the integrity of their pipelines by mandating periodic inspection and testing. Section 5 of S. 2004 would require internal inspections at least every 5 years; the Board does not have sufficient data to recommend a specific test period at this time. Section 5 of S. 2438 would require operators to periodically inspect and test pipelines; the Safety Board suggests that the frequency of the inspections or tests required depend upon the characteristics of the pipeline and the ability of inspection or test methods to detect defects before the defects propagate to critical size. Under such a mandate, RSPA could require that pipelines with protective coating deficiencies or known corrosion conditions be inspected more frequently, and that the frequency depend on the ability of the inspection method to detect defects.

The Safety Board supports language proposed in Sections 2b and 2c of S. 2409 that would require an operator to clearly define criteria for evaluating and acting on the results of inspections, and that would also require that prompt action be taken to address integrity issues.

The Safety Board has also urged, in safety recommendations, that pipeline employees be required to be trained, tested to assess the success of training, and periodically retrained and retested, as appropriate. Training requirements in Section 4 of S. 2438 are consistent with past Safety Board recommendations. Section 6 of S. 2004 would also require testing to determine whether individuals are qualified to perform assigned functions. Such a requirement is consistent with previous Safety Board recommendations. S. 2004 requires certification by the Secretary of Transportation, which the Board has not previously recommended. Objective criteria would provide regulators with specific means to reassess the qualification of individuals after an accident and before they resume regular duties.

Most of the provisions in Section 7 of S. 2438 are consistent with past Safety Board recommendations concerning public education and emergency preparedness needs.

Finally, the Safety Board is concerned that State officials' ability to inspect interstate pipeline operators is threatened. Effective oversight is needed to ensure that pipeline operators meet minimum standards. Section 9 of S. 2438 is consistent with the Safety Board's support for participation of State authorities in overseeing interstate pipeline activities.

Senator MURRAY. I would appreciate that, and Miss Showalter, you were asked whether you could oversee the interstate pipelines if that authority was given to you. All of a sudden the responsibility would be in your pocket if an accident like Bellingham occurred. Do you believe that there is, are the inspection divisions available today that the, that there is the ability right now to get the information you would need to assure citizens from Renton, Redmond, Bellevue, SeaTac that they have indeed safe pipelines that they were living next to?

Ms. SHOWALTER. You mean is there the technology available?

Senator MURRAY. Correct.

Ms. SHOWALTER. It's hard for me to answer that question. I'm not a technological expert. I can say that I think the State of Washington would have at least as good an ability as the Federal Government to avail itself of that technology, and as I said before, I think because we would put more resources into it and because I think because we're more responsive to the demand of our state we would do at least as good if not a better job to do that.

Senator MURRAY. Mr. Felder, would you respond to that?

Mr. FELDER. We want to say that we think the provisions in your bill on research and development and the need to strength research and development are exactly where they need to be. I've spoken to folks in the pipeline industry and outside of the pipeline industry, and quite frankly, I personally am astonished that as we're into the next millennium we do not have safety solutions from advanced technology to address the kind of problems that have been discussed in this room here today, and we would pledge to use additional funding to lead new research and development into advanced tools.

Senator MURRAY. I would absolutely agree with you, and my concern would be is if we do approve legislation at some point that gave Miss Showalter the authority to oversee the pipelines, is that there wouldn't be the inspections available to give you the assurance that you could give to the communities that they would be safe, and if we don't develop that research and development of testing, we would well have not made any progress whatsoever. So I will continue to work on that as well.

Mr. Chairman, I have other questions. If I could submit them for the record, I would appreciate it.

Senator GORTON. All right. I'm going to take one more crack at Mr. Felder.

Mr. Felder, you used the example of the Colonial Pipeline from Texas to New Jersey, I think quite appropriately, as an example of where obviously uniform Federal rules are vitally important. I don't know precisely where that pipeline goes, but the way I count my geography, it must go through 10 different states.

My question for Miss Showalter, however, was very different. The particular pipeline we're talking about here isn't 90 percent in the State of Washington. It's got to be 98 or 99 percent in the State of Washington. She said that if she were delegated or the state were delegated the authority, full regulatory authority over that pipeline, first, of course, she's here. I'm more directly responsive to the people of the state than any Federal entity can be, and second, that almost certainly the single state would have more, would have a greater number of inspectors. I asked her not about Colonial or a 10-state pipeline but whether or not it would be appropriate as a matter of the statute, the bill that Senator Murray and I, to make a statutory delegation to a state under those circumstances, i.e. 90 percent or more of a pipeline's length being within a single state if the state wished to accept that responsibility.

She gave me a quite enthusiastic yes to a question that she hadn't heard before. Is your answer to that question no? Would you and the administration oppose a mandatory delegation of authority under those circumstances?

Mr. FELDER. I can't, I can't really answer that question for the record, because I don't know the Administration's answer. I could give you a personal answer.

Senator GORTON. Yes, that's what I want then.

Mr. FELDER. I've been working in various forms of transportation in the public and private sector for my whole professional career, and I honestly do believe that the regulation of fixed facilities that operate in interstate commerce should be directed by the Federal Government. I think it's very important to work with the states, but I think when it comes down to it, if you're talking about the movement of goods and services around the country, I think the founding fathers got it right, that interstate commerce belongs in Federal hands.

Senator GORTON. No, the founding fathers said we had that authority if we wanted to take it. It doesn't require us to take it.

Mr. FELDER. No, it doesn't require us to take it. I couldn't agree with you more, but I'll just say that yes, I think, personally I think it should reside where it resides, but well informed by the Federal system.

Senator GORTON. OK. We perfectly understand your views. That's a straightforward answer to our question whether we agree with it or not.

I apologize to those of you that didn't get asked any questions during this period of time. We do value your testimony, and we value the testimony of all of you very much.

Thank you very much.

Senator GORTON. Well, the three of you have been waiting all afternoon for this opportunity, and we now wish to hear from you, and Mr. Gast, we'll start with you.

Mr. GAST. Thank you, Senator Gorton.

STATEMENT OF CARL GAST, MANAGER AND VICE PRESIDENT, OLYMPIC PIPE LINE COMPANY, RENTON, WASHINGTON

Mr. GAST. I'm Carl Gast, Manager and Vice President of the Olympic Pipe Line Company.

With me today is Tony Palagyi, Senior Environmental Project Manager, who has played a leading role with the Whatcom Falls Park restoration effort.

I'd like to thank Senators Gorton and Murray for inviting us to participate in this panel today.

Before I begin my remarks, allow me to introduce myself. I joined Olympic Pipe Line on January 3rd of this year. I oversee the day-to-day operations of the company here in Washington and in Oregon, as well as the implementation of the company's corridor safety action plan which is a comprehensive effort to address safety issues along the entire length of the pipeline from Ferndale to Portland.

As a certified engineer, I have 31 years of experience in the liquid fuels pipeline industry at various locations in the United States. Throughout my career as a pipeline engineer and manager, I've always made safety my No. 1 priority. I have been involved in a number of efforts to address safety issues for pipelines, but I must say that the safety action plan developed by Olympic in the last few months is the most far reaching effort with which I have ever been involved.

On behalf of myself and Olympic Pipe Line, I want to express again sincere sympathy and condolences to the families and friends of the three young people who lost their lives in a tragic accident. There is nothing I can say that will replace that loss, but I do want to talk about Olympic, what Olympic is doing to address safety issues along the entire pipeline, as well as our efforts to restore Whatcom Falls Park where the accident occurred.

In addition, I will touch upon our work with the Governor's Fuel Accident Prevention and Response Team and our community outreach activities along the pipeline corridor.

I'd like to start by mentioning the restoration efforts at Whatcom Falls Park. Olympic is working closely with members of the joint restoration committee, not only to restore Whatcom and Hanna Creeks, but to introduce significant improvements. It has been an excellent group effort. Proof of that significant effort is the return of salmon to the creek earlier than expected as well as the return of many native plant and animal species. Olympic will continue working closely with the state holders during this long-term restoration phase expected to last up to 5 years. Mr. Palagyi is here to answer your questions regarding the park restoration.

Next I'd like to discuss the two major safety action plans undertaken by Olympic. The Bellingham safety action plan was developed in cooperation with the City of Bellingham and addresses safety efforts in Whatcom and Skagit Counties. Our corridor safety action plan was approved in October by the board, Olympic board of directors as a comprehensive program to address safety along the entire pipeline. The actions we are implementing in these safety plans are intended to address all issues we understand are under investigation by the National Transportation Safety Board as a potential cause or contributing factor to the accident. We are not waiting until after the investigation is complete to take action. In other words, we are implementing safety actions in each area under investigation regardless of whether the area ultimately is found to have been involved in the accident. These areas include the integrity or condition of the pipe, valves, pressure levels, computer software and hardware, and the actions by employees who operate the pipeline from Olympic's control center in Renton, and in the time allowed me today, I would like to take a moment to discuss some of the safety actions Olympic is taking in the areas that we understand to be under this investigation. My written testimony provides a fuller description of all of the actions we are planning to take.

In the area of valves, Olympic has retained an independent professional engineering firm to determine whether changes should be made in the number, location and type of valves. The firm, Marmac, has completed its study of the northeastern segment of the line. As a result, Olympic has installed five new valves in the Bellingham area. Marmac is now working on similar studies of the pipeline all the way to Portland.

In regard to pressure issues, Olympic has retained an outside consultant to conduct computer simulated pressure surge analysis. The intent of these analyses is to show that the pressure might result at various points along the pipeline under a variety of operating conditions—normal, abnormal, and emergent—to insure that the pressure under these conditions will not exceed the pipeline's designed maximum allowable operating pressure or the maximum allowable surge pressure. These analyses are being done in concert with the valve placement study.

I should point out that the results of the surge analysis reenacted at the June 10th accident show that at the time of rupture of the pipe, the pressure in the pipeline at the rupture point was below its maximum allowable operating pressure and its maximum allowable surge pressure. In other words, the pipe failed at a pressure lower than what it was designed to withstand.

As many of you probably know, the NTSB has reported that the section of pipe that ruptured showed clear evidence of damage consistent with markings left by construction equipment. To check the condition of the line, the entire pipeline from Ferndale to Portland will be inspected internally.

CC Technologies of Dublin, Ohio has been retained to help develop and implement the internal inspection program. The entire pipeline will be inspected internally using two devices that employ the best accepted technology available today. One checks the roundness of the pipe. The other looks for areas of metal loss typically caused by corrosion or manufacturing defects. Olympic has voluntarily inspected its pipeline on about a 5-year basis, but the upcoming round of inspections is the most comprehensive program we have ever done.

In cooperation with the Office of Pipeline Safety and community representatives, Olympic will review the internal inspection data and determine if there are any anomalies requiring visual inspection. In cooperation with the Office of Pipeline Safety and community representatives, Olympic will review the internal inspection data and determine if a visual inspection of the pipe is required. Olympic is also a committed partner to the state "one call" program which requires excavators to call before they dig near utility lines to give utilities opportunities to properly mark their lines and/or observe any digging around their lines.

I want to emphasize that our safety action plans are not an end in themselves but an ongoing continuous effort as we address safety along the entire pipeline.

That outlines some of the actions we have taken that support our January 14th request to the Federal Office of Pipeline Safety for permission to restart flow of product through the segment of 16-inch line that's been shut down since June 10th.

At this point we do not know when the Office of Pipeline Safety will complete its review of all of the materials we have submitted and make its decision regarding restart.

I'd like to take a few moments now to discuss our recent work with the Governor Locke's Fuel Accident Prevention Response Team. We want to acknowledge the hard work of the team. We appreciate their recommendations to approve public awareness and strengthen emergency response. We also support state efforts to strength our one call system.

Our issue of current debate is the degree of authority the state would allow over interstate pipelines. In our view, the existing division of regulatory responsibility makes sense. We are concerned about the potential for patchwork regulations that differ from state to state and the implications it would have on interstate commerce. Having a unified set of regulations is important for smooth regulation. Ultimately, we recognize that the balance between Federal and state authorities is up to elected officials such as yourself. However, we encourage you to consider carefully the need for a unified set of regulations for interstate pipelines.

Last, certainly not least, I'd like the talk about the community involvement. Olympic is committed to reaching out to communities along the pipeline. Our corridor safety action plan contains a strong community out reach core. Since the board approved this plan last October, we have met with local media up and down the corridor as well as held three major community briefings in December in the three most populace communities through which the pipeline travels. We have held numerous meetings with local city councils, other elected officials, local emergency response groups, school districts, and neighborhood locations.

As I mentioned before, we are also committed to reviewing with the communities along the pipeline the data from our upcoming internal and field inspections, our valve detect study, and our surge analysis. In fact, tomorrow, Wednesday, and Thursday of this week we are holding three pipeline integrity workshops to help technical community representatives better understand the results of our internal inspections as we move ahead with the corridor safety action plan.

In conclusion, I'd like to stress that Olympic's safety action plans are comprehensive and community oriented. We are dedicated to working with communities and elected officials to address safety along the entire pipeline. We are working closely with the Governor's response team, the state legislature and others in pursuit of a common goal. We believe that the Whatcom Falls Park restoration effort is an example of what we can achieve if we work together.

If you have any questions, I'd be happy to answer them for you now or later.

[The prepared statement of Mr. Gast follows:]

PREPARED STATEMENT OF CARL GAST, MANAGER AND VICE PRESIDENT, OLYMPIC PIPE LINE COMPANY, RENTON, WASHINGTON

I am Carl Gast, manager and vice president of the Olympic Pipe Line Company. With me today is Tony Palagyi, Senior Environmental Project Manager, who has played a leading role in the Whatcom Falls Park restoration efforts. I would like to thank Senator Gorton and members of the Senate Commerce, Science and Transportation Committee for inviting us to participate in today's panel.

Before I begin my remarks, allow me to introduce myself. I joined Olympic on January 3rd. I oversee the day-to-day operations of the company here in Washington and Oregon, as well as the implementation of the company's Corridor Safety Action Plan, which is a comprehensive effort to address safety issues along the entire length of the pipeline from Ferndale to Portland.

As a certified engineer, I have 31 years of experience in the liquid fuels pipeline industry at various locations in the United States. Throughout my career as a pipeline engineer and manager, I have always made safety a priority. I have been involved in a number of efforts to address safety issues for pipelines, but I must say that the Safety Action Plan developed by Olympic in the past few months is the most far-reaching effort with which I have been involved.

On behalf of myself and Olympic Pipe Line Company, I want to express again sincere sympathy and condolences to the family and friends of the three young people who died in that tragic accident. There is nothing I can say that will replace the loss, but I do want to talk about what Olympic is doing to address safety issues along the entire pipeline, as well as our efforts to restore Whatcom Falls Park, where the accident occurred. In addition, I will touch upon our work with the governor's Fuel Accident Prevention and Response Team, and our community outreach activities along the pipeline corridor.

Whatcom Falls Park Restoration

I would like to start by discussing restoration efforts in Whatcom Falls Park. Olympic Pipe Line Company is working closely with members of the Joint Restoration Committee not only to restore Whatcom and Hannah Creeks, but also to introduce significant improvements. The Joint Restoration Committee is composed of representatives and specialists from Olympic, the City of Bellingham and other consultants. They are advised by the Trustees which includes a committee composed of the Department of Ecology, Department of Natural Resources, National Oceanic and Atmospheric Administration, Lummi and Nooksack Tribes, and the Department of Fish and Wildlife.

The initial funding for all the restoration and improvement efforts has been provided by Olympic Pipe Line Company.

• **Restoration.** Creek banks and plant life are being restored by the collaborative efforts of the Restoration Committee and nature. Here are some examples:

1. Fallen trees were added along the bank to reduce perennial flooding of both Whatcom and Hanna Creeks.
2. Woven matting was used extensively at Hannah Creek to solidify the banks against erosion. Matting holds the soil while allowing grasses and plants to grow through the mesh.
3. By October, plants and grass were growing up through the matting and a significant amount of native plants, such as ferns and salal, were making a comeback. In addition, invasive plants such as Canadian thistle and reed canary grass are being removed.
4. Hydromulching was added to creekbanks to reduce erosion, while native grass seed was added to the banks for soil retention.
5. Trees in the burn zone will be monitored through the spring of 2000 to determine which scorched trees will recover. More trees and shrubs will be planted this spring.

• **Improvements.** In addition to restoration efforts on the creek, Olympic has worked with the Joint Restoration Committee team to improve significantly the creeks and the surrounding affected area. Improvements include:

1. Building 30 new spawning pools for the salmon.
2. Creating pools and minimizing obstructions to help salmon move upstream.
3. Adding creek meanders and placing logs in the creek banks to help retain the banks of the streams. The increase in meanders helps slow Hannah Creek down, which improves the fish habitat.
4. Establishing intentional log/debris spots in the creek. This builds habitat needed to capture organic matter that nurtures and sustains the life of the creek at the micro level. The logs also provide refuge for fish.
5. A total of nearly 6,000 pounds of metal debris was pulled from the creek during the clean up.
6. We are replacing the Valencia Street Bridge and making improvements directly related to the park.

• **Salmon are returning.** As early as late August, salmon were returning and had made it as far as naturally possible to Pixie Falls, inside Whatcom Falls Park. And late November, approximately 20 Chum Salmon were sighted at the Falls, which is more than usual and is considered a good sign.

Ultimately, restoration efforts have increased salmon habitat in the park by about 60 percent. Along with these restoration efforts, creek improvements will enhance the habitat needed by salmon and resident fish such as rainbow and cutthroat trout for years to come.

• **Long-Term Restoration:** Olympic will continue working closely with the Natural Resources Trustee group, chaired by Clare Fogelsong of the City of Bellingham, and the Department of Ecology. The long-term restoration phase expected to last up to five years.

Key elements of the Long-Term Restoration effort will include:

1. Control of invasive plants
2. Re-vegetation
3. In-stream habitat improvements
4. Multi-year monitoring
5. Potential land acquisitions and,
6. Monitoring ground water.

Olympic's Safety Action Plans

Next, I would like to discuss two major safety action plans undertaken by Olympic. The Bellingham Immediate Safety Action Plan was developed in cooperation with the City of Bellingham and addresses safety efforts in Whatcom and Skagit Counties. Our Corridor Safety Action Plan was approved in October by the Olympic Board of Directors as a comprehensive program to address safety along the entire pipeline.

Following the June accident, the Office of Pipeline Safety issued a Corrective Action Order, which it amended twice. These directives spelled out the steps Olympic was required to take before receiving permission to resume operation of the pipeline.

Collectively, these safety actions address all the issues we understand are under investigation by the National Transportation Safety Board as potential causes or contributing factors to the accident.

These areas include the integrity or condition of the pipe, valves, pressure levels, computer software and hardware and actions by the employees who operate the pipeline from Olympic's Control Center in Renton.

In other words, we are implementing safety actions in each area under investigation, regardless of whether that area ultimately is found to have been involved in the accident.

Here are some of the major actions we have taken to comply with the requirements of the federal Office of Pipeline Safety.

1. Valves

First, investigators are attempting to determine whether a pressure relief valve may have failed to function properly, and if that was a factor in the June 10 accident. Investigators are also evaluating whether the number of mainline block and check valves could have reduced the size of the release.

Here's what OPL is doing or has done:

- OPL has tested its relief valves and commissioned a detailed engineering study of the delivery facility where the relief valve was located.
- Olympic also has retained an independent professional engineering firm to determine whether changes should be made in the number and location of valves to reduce the potential impact of an accidental release on environmentally sensitive areas and population centers.
- Marmac has completed a study of the northern segment of the line, and as a result, Olympic has installed five new valves in the Bellingham area.
- Marmac is now working on similar studies of the pipeline all the way to Portland.
- In addition, it's important to note that Olympic regularly tests the pipeline mainline valves and will perform additional tests in the upcoming months. These tests are conducted through field inspections by an operations technician as well as through monitoring at its Control Center in Renton.

2. Pressure Issues

Second, investigators are attempting to determine if a pressure increase in the pipe contributed to the accident.

Here's what OPL is doing or has done in this area:

- **Pressure Surge Analysis:** Olympic has hired Stoner Associates to conduct computer simulated pressure surge analyses to show what pressures might result at various points along the pipeline under a variety of operating conditions.
- **June 10 re-enactment.** Results of the surge analysis re-enacting the June 10 accident showed that, at the time, the pressure in the pipe at the rupture point was below both its maximum allowable operating pressure and its maximum allowable surge pressure, as established under Department of Transportation regulations.

3. Line Integrity

The third issue investigators are looking into as a possible cause of the rupture is line integrity or the condition of the pipeline. The NTSB has reported that the section of pipe that ruptured showed clear evidence of damage consistent with markings left by construction equipment.

Here's what Olympic is doing about the issue of line integrity:

• **Internal Inspections:** The entire pipeline from Ferndale to Portland will be inspected internally. CC Technologies of Dublin, Ohio, has been retained to help develop and implement the internal inspection program. This includes:

- Identifying and evaluating available inspection tools;
- Developing a process to use and verify the information;
- Developing a method and protocol for inspections;
- Implementing the program systemwide.

Olympic likely will begin the process of internal inspection in Whatcom and Skagit Counties. The section of the pipeline from Ferndale Station in Whatcom County to Allen Station in Skagit County will be inspected using three state-of-the-art devices: a geometry device; a high resolution magnetic flux device; and an ultrasonic device. These inspection devices run inside the pipe and are propelled down the pipeline by the flow of petroleum products.

The Office of Pipeline Safety must approve the selection of the devices that will be used for the internal inspection of the pipeline.

Each device is designed for a specific purpose.

• A *geometry device* looks for changes in the roundness of the pipe.

• A *high-resolution magnetic flux device* uses a magnetic field to locate and identify metal loss due to internal or external corrosion.

• An *ultrasonic device* utilizes ultrasonic pulses to inspect the pipe. This device is designed to identify the same range of features and defects as the high-resolution magnetic flux tool but uses ultrasonic pulses rather than a magnetic field to collect the data.

• **Field Inspections and Repairs:** In cooperation with the Office of Pipeline Safety and community representatives, Olympic will review the internal inspection data and determine if there are any anomalies requiring visual inspection. In cooperation with the Office of Pipeline Safety and community representatives, Olympic will review the internal inspection data and determine if a visual inspection of the pipe is required. These field inspections will be undertaken based on Department of Transportation requirements and any repairs will be made in accordance with standards set by the American Society of Mechanical Engineers.

Olympic plans to add another step to the internal inspection process. The Company will dig up and visually inspect an additional number of anomalies that fall below the criteria for excavation as approved by the Office of Pipeline Safety. The purpose of this additional step is to compare the internal inspection data with a visual inspection of the corresponding portion of pipe.

Olympic also is a committed partner in our state's One-Call Program which requires excavators to call before they dig near utility lines to give utilities the opportunity to properly mark their lines and/or observe any digging near or around their lines.

4. Operations Controller Actions

Investigators are evaluating whether any actions taken on June 10th by the employees who operate Olympic's computer system in Renton were a contributing factor to the accident.

Here are the actions OPL has undertaken:

• **Operator Re-Training:** Olympic has completed a re-training program for its operations controllers, who operate the pipeline from Olympic's Control Center in Renton. Training also is being provided for the technicians who perform a variety of field tasks, including maintaining pumps and valves, testing the petroleum stream, and supervising the "one-call system" in effect when anyone digs near the pipeline. Technicians who cover Whatcom and Skagit Counties have already received additional training.

The retraining program goes beyond requirements of the Office of Pipeline Safety's first amended corrective action order in that it includes training for the technicians. The program also begins the process of early fulfillment of OPS' "final rule" announced August 26 that requires pipeline operators to develop and maintain a written qualification program for individuals performing safety tasks on pipeline facilities.

All employees in other areas along the corridor and throughout the company will receive training and be re-qualified by the end of next year, well in advance of the deadline of Oct. 1, 2002 set by OPS in its final rule.

5. Computer software and hardware

Investigators are attempting to determine whether an internal database error, along with a simultaneous increase in processing demands, caused a computer slowdown on June 10th. During the computer slowdown, the controllers were unable to obtain current pipeline information on the computer screens and to process commands to equipment, such as pumps along the pipeline.

Here's what Olympic has done in this area:

- **Computer Analysis and Upgrade:** Olympic has completed an analysis of its Supervisory Control and Data Acquisition System (SCADA), the software that operates its computer system. Based on that analysis, system parameters have been modified. Olympic has also made modifications, upgrades and design changes to its computer system, including an increase of 750 percent in processing capacity.

That outlines some of the actions we have taken that support our January 14 request to the federal Office of Pipeline Safety for permission to restart the flow of product through the segment of our 16-inch line that has been shut down since June 10.

At this point we do not know when the Office of Pipeline Safety will complete its review of all the materials we have submitted and make its decision regarding restart.

However, I would like to review with you what Olympic will do once we receive written authorization to resume flow of product.

First, personnel trained in startup procedures will be present at the Ferndale and Bayview stations during the first 12 hours of operation. Trained personnel also will be stationed at block valves at milepost 7 (north of Bellingham) and milepost 16 (at Lakeway Drive in Bellingham).

Two controllers with a minimum of ten years' operating experience will be on duty at Olympic's control center in Renton at all times during the startup.

Only diesel and jet fuel, which are less volatile than gasoline, will run through the pipeline during the first week of operation.

The pressure in the line running from Ferndale to Bayview will be limited to 70 percent of its normal operating pressure until it is verified that the system is operating properly. At that time, pressure may be increased only to 80 percent of its normal operating pressure. The pressure in the entire system is limited to 80 percent until Olympic receives written authorization to increase the pressure from the Office of Pipeline Safety.

The successful startup will signal the next phase of testing we have agreed to do that will include the use of different internal inspection devices throughout the entire system, as previously described.

I want to emphasize that our Safety Action Plans are not an end in themselves, but an on-going continuous effort as we address safety along the entire pipeline.

Governor's Fuel Accident Prevention and Response Team

I now would like to take a few moments to describe our recent work with Governor Locke's Fuel Accident Prevention and Response Team. I want to acknowledge the hard work of the Team. Pipeline safety covers a broad range of issues, from complicated legal questions of federal, state and local government relationships, to highly technical engineering questions of pipeline operations, as well as complex logistical questions of emergency response. We appreciated the Team's efforts to begin to tackle the issues presented.

The Team made several recommendations concerning public awareness, local emergency response, and the One-Call system, many of which are now the subject of pending state legislation.

Olympic believes these issues are all extremely important, and supports recommendations that would strengthen and improve public awareness and emergency response.

Olympic strongly supports continued development of the One-Call system. One statewide number will reduce confusion.

One issue that keeps coming up is the degree of authority states will have over interstate pipelines. In our view, the existing division of regulatory responsibility makes sense. Further, we are concerned about the potential for patchwork regulations that differ from state-to-state and the implications that would have for interstate commerce. Having a unified set of regulations is important to its smooth operation.

Ultimately, we recognize that the balance in this issue is up to elected officials such as yourselves. However, we encourage you to consider carefully the need for a unified set of regulations for interstate pipelines.

Community Outreach

Last but certainly not least, Olympic is dedicated to reaching out to the communities along the pipeline. Our Corridor Safety Action Plan contains a strong community outreach core. Since the Board approved this plan last October, we have met with local media up and down the corridor, as well as held three major community briefings in December in the three most populous counties through which our pipeline travels. We have held 64 meetings with local governments, other elected officials, local emergency response groups, school districts and neighborhood associations.

We also are committed to reviewing with the communities along the pipeline the results from our upcoming internal and field inspections, our valve effectiveness study and our surge analyses.

Since the new inspections won't be completed for a number of months, many communities have asked us to review the results of our 1996/97 inspections. However, our recent experience in sharing this technical information with some communities is that it can be difficult to understand.

Therefore, this week we are holding three Pipeline Integrity Workshops that are designed to assist communities in understanding the 1996/97 data as well as our upcoming inspections, and the criteria used to conduct field inspections. The workshops also will cover how the technology employed by the inspection tools we are using in 2000 differs from what was used in 1996/97. Our consultant, CC Technologies will be conducting the workshops.

Conclusion

In conclusion, I would like to stress that Olympic's Safety Actions Plans are comprehensive and community-oriented. We are dedicated to working with communities and elected officials, such as yourselves, to address safety along the length of our pipeline. We have worked closely with the Governor's Response Team, the state legislature, and others in the pursuit of this common goal. We believe the Whatcom Falls Park restoration effort is an example of what we can achieve if we work together.

Now, if there are any questions I will be happy to answer them now or in writing. Thank you.

Senator GORTON. We'll defer most of our questions until we've heard from the other two witnesses, but I do have one for you now. Where is your predecessor, the person that was in charge on June 10th?

Mr. GAST. He is in Houston, Texas.

Senator GORTON. OK. Mr. Sluder.

STATEMENT OF ROBERT L. SLUDER, VICE PRESIDENT, WILLIAMS GAS PIPELINE-WEST, SALT LAKE CITY, UTAH

Mr. SLUDER. Thank you, Senators Gorton and Murray. I'm Robert L. Sluder, Vice President of Operations for Williams Gas Pipeline-West. Williams is a diversified energy and communications company with operations in 50 states.

Williams gas pipeline unit consists of five wholly owned interstate pipeline systems. In aggregate, we move approximately 16 percent of the natural gas consumed in the United States.

Senator GORTON. Natural gas only? No liquid?

Mr. SLUDER. Natural gas. Our western operations include Northwest Pipeline which runs from New Mexico through the states of Utah, Colorado, Wyoming, Idaho, Oregon and Washington to the Canadian border. We operate approximately 1,400 miles of pipe in the State of Washington and have done so since the pipeline was constructed in the late 1950's. PG&E gas transmission northwest also operates the other interstate pipeline in the State of Washington. Together we deliver virtually all of the natural gas that is consumed in the State of Washington. We do that through our four local, four major customers who are local distribution company or intrastate companies. Those are Cascade Natural Gas, Puget Sound Energy, Avista and Northwest Natural. We also have direct connection to several industries and municipalities in the State of Washington.

Williams is testifying today not because we were directly involved in the incidents here in Bellingham, but because we have been drawn into the debate of overall pipeline safety as a result of those incidents, and we welcome the opportunity to participate in this debate.

We understand and accept the fact that Federal, state and local governments want to be involved in our industry safety efforts, so that they can assure the public that we are providing a safe service.

I'd like to state or make the assertion that states already play an important role in that safety effort. It's been mentioned earlier that the intrastate system in the United States is under the governance and the jurisdiction of the states. That amounts to 75 percent of the total pipeline mileage in this country.

Second, the states' "one call" laws are a major factor in interstate pipeline safety, and last, state inspectors accompany DOT personnel on audits throughout all of our system. Whether those states are authorized as was discussed earlier by DOT or not, it is my experience that they have always been welcome by the DOT and have accompanied the DOT inspectors on many inspections in the states of Utah, Colorado, et cetera. We strongly believe, however, that for interstate pipeline systems, the Federal Government is the appropriate agency to set standards and to administer interstate programs. We believe that the interstate natural gas transportation industry has a good safety record.

Attached to my formal testimony are statistics that I believe will bear that out. We have achieved that record not by chance. Safety for us begins at the mill where the steel is rolled into the pipe and very strenuous metallurgical testing is done not only to the pipe, but to the valves, the fittings and all of the other appurtenances that go into our system. Through careful design and construction the pipelines are built and constructed and ultimately hydrostatically tested with water prior to putting them into service.

Now, in theory, if nothing changed after that initial hydrostatic test, there would be no reason to doubt the strength of the steel and the concept is it may last forever, but things do change. The soils attack the pipe, the coating. Land forces attack the steel, itself, and exert pressures on the pipe. External digging occurs around the line. So many things do change, but we don't sit by as was suggested earlier and do nothing. We do many things to observe the changes and monitor them and make decisions about safety and maintenance in conjunction with those. We routinely fly our rights-of-way and in many congested areas in cities and towns we drive or physically walk those rights-of-way. We look for signs of unusual activity, either unusual construction activity or unusual activity such as landslides or other stream erosion type events. We participate very strongly with the states in the "one call" program which is designed to prevent the unintended digging near our pipelines. We also monitor our cathodic protection systems designed to mitigate the effects of corrosion. We assess all of these concerns and in many instances then make the decision to internally inspect the pipeline.

We do run smart pigs. We have run smart pigs in many locations in the State of Washington on the Northwest Pipeline system, but we did so after it was warranted by the information that I described earlier.

Changes in population occur along our pipeline. The pipeline again was built in the 1950's. Many areas have developed and grown up around the pipeline. DOT requirements have for many years required us to do either of several things, either reduce the pressure that we operate the pipeline at when an increase in population has grown up along the pipeline, which is really not practical, because we are obligated to provide a service and our ability to move gas is a function of the pressure that we operate under, and so traditionally what we've done is go in and replace the pipe. Dig it out and replaced the pipe with heavier walled pipe in heavier density area. Today we look at that, and we also look at smart pigging the line, internally inspecting the line and not removing good quality pipe if we can determine that it is good quality pipe, and often times we'll do hydrostatic testing in those areas as well. What we have to do is judge what inspection tools and practices would be most effective at any given location in the pipeline.

Lastly, to ensure safety, we work with public education programs. We provide the residents and emergency response officials all along this system the information about the operations and appropriate actions to take during normal operations and in the event of an emergency.

And then there are unique circumstances. In the State of Washington we are susceptible to landslides, not only Washington, we have the same circumstances through parts of Oregon, Idaho and Colorado. Utah as well. To cope with that we have instituted a special monitoring program wherein we have identified over 95 active landslides and have instrumentation in those landslides to protect the pipeline, and I hope that the community can see that it's not just one-size-fits-all in protecting pipelines.

As important as safety is, there are two other factors that are very important as well, and very important to interstate commerce, and that is the reliability of our service and the affordability of our service.

Interstate pipelines operate as an integrated system. Our Northwest Pipeline has natural gas entering the system at three locations, on the extreme southern end in the states of Colorado and New Mexico, in the mid-section in the Rocky Mountains of Idaho, Utah and Wyoming, then gas that is brought across from Canada and produced in B.C. and Alberta. Gas moves through local distribution customers and end users in a variety of ways, and it changes as pricing patterns change, and as loads on their systems change.

The question came earlier about state authority. Mr. Felder gave an example of a liquids line in the east. I can give an example right here in the State of Washington that is perhaps even more pertinent. If Idaho decided to increase the safety margins by requiring that pipelines operate at lower pressures in their state, our ability to deliver gas to Oregon and Washington would be adversely affected by virtue of the fact that most of the gas that comes into this state comes from the south, not all of it comes from the Canadian side. If on the other hand Washington chose to mandate hydrostatic testing, whole lines would have to be taken out of service, and customers south of this state, i.e., Oregon, Idaho, et cetera, would suffer the effect of that service disruption.

Under Federal jurisdiction there is consistency in interpretation and application of regulations and a consideration for the overall interstate commerce that we are involved in.

Again, I do not imply that states do not have a role to play. I believe they play an important role, again, by virtue of their authority over the approximate one and a half million miles of intrastate pipelines. They play a major role in that system.

Secondly, their "one call" systems, the statistics indicate that 80 percent of those accidents involving death or injuries can be attributable to dig-ins, and it is our commitment in the pipeline industry to work hand and hand with the states and to seek better laws and to seek better enforcement of those laws to promote better safety.

In conclusion, the interstate natural gas pipeline systems are proud of their safety record, and I personally am very proud of the Washington-based employees that our company employs here and that work day-in, day-out to provide the natural gas service and do so in a safe, reliable and efficient manner. We certainly are actively seeking ways to improve on that record. However, I cannot promise that there will never be another pipeline incident, but I can promise that we are committed to take every effective action we can to prevent the next one from ever happening.

We are also committed to working with the Office of Pipeline Safety as they do their job of overseeing the safety efforts of our interstate industry. We hope that Congress will give us and the Office of Pipeline Safety the flexibility we need to accomplish this mission.

Thank you.

[The prepared statement of Mr. Sluder follows:]

PREPARED STATEMENT OF ROBERT L. SLUDER, VICE PRESIDENT, WILLIAMS GAS PIPELINE-WEST, SALT LAKE CITY, UTAH

INTRODUCTION

Mr. Chairman and Members of the Committee, my name is Robert L. Sluder. I am Vice President of Operations for Williams Gas Pipeline—West. I appreciate this opportunity to appear before the Committee on behalf of Williams to discuss natural gas pipeline safety issues.

Williams is a diversified energy and communications company with operations in all 50 states. Our Williams Gas Pipeline unit consists of five wholly owned interstate natural gas pipeline systems. We also have partial ownership in several other systems. Approximately 16% of all natural gas consumed in the United States is transported on Williams' system.

Our western operations include Northwest Pipeline, which runs from the New Mexico/Colorado border, through Utah, Colorado, Wyoming, Idaho, Oregon and Washington to the Canadian border. We operate 1,400 miles of pipe in Washington and have done so since the system was constructed in the late 1950s. PG&E Gas

Transmission—NW also operates an interstate pipeline in Washington. Virtually all of the natural gas consumed in the state is delivered through these two pipeline systems. Our primary customers are the four local distribution companies operating in Washington—Cascade Natural Gas, Puget Sound Energy, Avista and Northwest Natural. We also deliver gas directly to some industrials and municipalities. Attached to my testimony is a map, which shows the Northwest and PG&E Gas Transmission systems and the service territories of the local distribution companies we serve. Williams has recently proposed to construct a new line from our Sumas, Washington, location across Whatcom County to Vancouver Island.

SUMMARY

At the outset, let me summarize my testimony. The interstate natural gas pipeline industry has an excellent safety record—superior to other modes of transporting fuel or products. Operating and maintaining a safe system is embedded in every aspect of our business. The cost associated with having an accident—both human and financial—far outweigh the costs of a responsible safety program.

We understand and accept the fact that Federal, state and local governments want to oversee the industry's safety efforts so that the public can be assured that we are making every reasonable effort to insure safety. States already play an important role in insuring the safe delivery of natural gas by virtue of their jurisdiction over intrastate pipelines, which comprise 75% of the total pipeline mileage in the country. States also have jurisdiction over "one-call" or "call-before-you-dig" programs. Since 80% of the accidents on intrastate natural gas pipelines involve third party digging activity, the strength of a state's one-call laws is a major factor in interstate pipeline safety.

We strongly believe, however, that for interstate pipeline systems, the Federal government is the appropriate body to be the safety regulator. Our systems cross state boundaries. Our operations and safety activities are planned and implemented on a system-wide basis. If we faced separate regulations in every state in which we operate, we believe the net result would be a less efficient, less safe system overall.

NATURAL GAS AND OUR SAFETY RECORD

For natural gas pipelines, there is no more important objective than to operate and maintain a safe system. It is in our own best interest to do so. Natural gas transmission pipelines have an excellent safety record. Attached to my testimony are two charts. The first shows the safety record of natural gas pipelines compared to other forms of transportation. The second shows the number of incidents, injuries, and fatalities for natural gas pipelines over the last ten years. While this record encourages us, we continuously seek ways to improve our current practices and seek new technologies to further enhance safety.

Accidents have serious and unacceptable consequences. Accidents can cause injuries or death, to our own employees and the public. They also can disrupt service to our customers and limit the full utilization of our systems for an indefinite period of time. Accidents cause the public and government officials to question our commitment to safety and the credibility of the Office of Pipeline Safety (OPS). The recent accident here in Bellingham demonstrates this quite clearly.

Unless the public believes that the industry can produce, transport and distribute natural gas safely, we cannot stay in business, and America would lose its cleanest fossil fuel.

I should note that the physical differences between natural gas and petroleum products are such that the characteristics of a natural gas pipeline accident differ from that of a petroleum pipeline accident. Natural gas is primarily composed of methane gas. Natural gas is lighter than air, insoluble in water and does not cause environmental damage when released. Most of the incidents that occur in natural gas transmission pipelines are small leaks. These leaks are easily found by pipeline personnel during their routine inspections and are fixed. In a worst-case scenario—a total pipeline rupture—only the area directly around the rupture presents a direct risk to the public at the initiation of the rupture. The size of this area depends on the pressure and size of the line but generally should not extend more than 600-feet on either side of the pipeline rupture. After the initial rupture, the noise of the escaping natural gas or, if there is ignition, the resulting fire will cause people to move away from the location immediately.

The probability of such an event occurring along the interstate natural gas transmission pipelines in the U.S. is small. During the period of 1986–1999, onshore interstate gas transmission accidents have caused no fatalities and nine injuries among the general public. This is a commendable record for a system that transports over one quarter of the nation's energy needs.

STEPS WE TAKE TO ASSURE SAFETY

A natural gas pipeline approaches safety from a system-wide perspective. Pipelines implement and comply with the minimum safety standards imposed by the Office of Pipeline Safety. Williams, along with other interstate pipelines, has internal procedures that exceed these minimum requirements in many respects.

The first step in the safety process is to make sure that the line is constructed properly. Safety actually begins at the mill where the steel is made. Pipeline representatives inspect the pipe at the mill to insure that it meets quality standards. During construction, the integrity of coatings designed to protect against corrosion are checked and imperfections are corrected. Welds are quality checked with x-rays. The line is tested with water to a pressure significantly higher than its maximum operating pressure prior to going into service.

Once a pipeline is in the ground, it is monitored in a number of ways. Among the procedures we employ, employees physically walk and inspect the pipeline corridor periodically. Companies also fly the right-of-way at least once a month, often more. In both cases, we look for signs of unusual activity on the right-of-way or any discoloration of plants or grasses that might indicate a small leak. Companies participate in one-call programs designed to prevent unintended digging near pipelines and other underground facilities. Employees test for leaks using analyzers and verify the effectiveness of cathodic protection, electrical systems that prevent corrosion on a pipeline. Valves are inspected and compressor engines are maintained. Any missing pipeline markers used to identify the location of the pipeline are replaced. In areas where we suspect corrosion may have degraded the integrity of the pipe, we do periodic internal inspections utilizing specialized detection equipment commonly known as "smart pigs."

Specific measures are incorporated into the regulations to raise the level of safety of natural gas transmission pipelines as the population density around our pipeline increases. These categories of population density, known as class locations, range from rural (Class 1) to heavy urban (Classes 3 and 4). As the population density increases, more stringent design, construction, inspection and maintenance practices are stipulated. We are required to walk our entire system once a year to monitor the area around the pipeline for changes in population density. When these changes occur, the pipeline operator is required to insure that the installed pipeline meets the criteria for pipe design that applies in the higher class location. If it does not meet these requirements, the pipe is replaced, the operating pressure in the line is reduced, or similar safety measures are undertaken to achieve the required margin of safety. The new class location also requires increased frequency of many different inspections.

A pipeline operator also gains a tremendous amount of knowledge about the condition of the line as systems are expanded, new meter stations or delivery points added, and laterals attached. While these activities are not directly safety related, they involve digging up parts of the system and documenting the condition of the pipeline, thereby giving the operator additional information to assess the integrity of the pipeline.

All of the information that a company gathers about its system goes back into tailoring the safety activities of the company, so that parts of the system in the greatest need of attention receive greater scrutiny. For example, we decide where and when to run smart pigs based on this accumulated knowledge. Smart pigs are very good at providing certain types of information about the condition of the pipe, but they do not provide a complete solution. We have to judge what inspection tools and practices will be the most effective at any given location on the pipeline.

Federal law requires pipelines to have public education programs. We provide residents who live along our rights-of-way with information about the pipeline, including what activities to look for and what to do in an emergency. We provide information and our emergency phone number to call in the event of seeing anything unusual. We work with local emergency response officials to educate them about the nature of our operations and the appropriate actions to take if there is an accident. We encourage our employees to interact with local officials and educate them about the pipeline. Unfortunately, our experience often is that emergency response personnel and local officials are often so pressed by the immediate demands facing them that getting their attention to learn about pipeline facilities that have never caused them any problems can be difficult.

The level of safety effort is substantial. A recent survey by our trade association, the Interstate Natural Gas Association of America (INGAA) revealed that its pipelines members, operating 160,000 miles of pipe, spent about $560 million a year, or about $3,515 per mile on safety related efforts. Our transmission pipeline industry, since the 1950s, has supported and been active in two organizations that research ways to improve safety practices and technology—the Gas Research Institute (GRI), and the PRCI (Pipeline Research Committee International). These organizations budget millions of dollars each year for these activities. In addition, since 1990, the INGAA Foundation has supported the efforts of companies who develop and promote safety technology in the marketplace.

Without question, natural gas pipelines are committed to safety and have consistently demonstrated a commitment to invest substantial sums to maintain and protect their systems and the surrounding environments. However, conditions differ from system to system and from location to location on a given system, making it difficult to create one-size-fits-all rules for when each activity is performed. Lines in damp soil will require a different type of attention than lines in the desert. Lines in areas where coal mining has occurred are susceptible to subsidence problems, whereas lines in other areas are not. In our case, we have found that at certain hillside locations here in Washington, the ground has become unstable after periods of uncharacteristically prolonged rainfall. We have experienced landslides that pulled the steel pipeline apart. We now have instituted a special monitoring program identifying and targeting these locations, but it is a localized problem, not system-wide.

My point is that the details of an effective safety program will vary from pipeline to pipeline and even on a given system. An effective program will focus resources differently from year to year, depending on the needs of the system.

REACTION TO WASHINGTON STATE ACTIONS

Williams made presentations to the Governor's task force during its deliberations, and I have spoken before the committees of the state legislature implementing the Task Force recommendations. I do not want to repeat all of that testimony here, but there are two key areas I believe it would be useful to address in the context of this hearing.

• National Standards vs. State Authority

The report from the Governor's task force raises the issue of whether to retain exclusive Federal jurisdiction over interstate natural gas and hazardous liquid lines or to give states authority in this area as well. It is our strong conviction that retaining exclusive Federal regulation of interstate pipeline safety is critical. Let me give you several reasons why we believe continuing the current structure will benefit the public, both from the standpoint of safety and service.

As I mentioned, pipelines are operated as integrated systems. Our Northwest pipeline has natural gas entering the system at three locations: gas produced from the San Juan basin of Colorado and New Mexico; gas produced from the overthrust belt area of the Rocky Mountains, and gas produced in Canada. Gas moves to local distribution customers, end users, or other pipeline customers in a variety of ways that change as pricing patterns change. We have designed a system that allows this to occur without regard to the weather or other factors that influence demand for natural gas. Safety work is scheduled so as not to interfere with the basic operation of the system.

If states have the authority to impose more stringent safety standards, they could interfere with operational flexibility and thereby impact our ability to serve customers, including customers in other states. For example, if Idaho decided to increase the safety margin by requiring that pipelines operate at lower pressures, our ability to deliver gas to Oregon and Washington would be adversely affected. If a state were to require hydrostatic testing, whole lines would have to be taken out of service, wreaking havoc on our ability to serve. Every time a state would adopt a new or different testing or inspection requirement, it would reduce an operator's flexibility and ability to operate the system according to the needs of the system as a whole.

Allowing states to impose different safety standards would also complicate the compliance process. Generally, safety teams move up and down the system performing tests and maintenance. In fact, the industry has moved more toward this functional approach in order to achieve greater uniformity of safety programs across the systems. If these individuals have to apply different standards in different states, it will erode the very uniformity we are trying to achieve.

Just as it makes sense for pipelines to adopt a holistic approach to safety, oversight by inspectors should take the same approach. When the Office of Pipeline Safety conducts an inspection, they are not limited to examining our practices within a single state, but can look system-wide. Also, Federal enforcement allows for consistency in the interpretation and application of regulation. If we get into a situation where different state inspectors interpret regulations differently, the resulting confusion will hurt, not help, safety.

Finally, if state actions force a company to allocate its safety dollars in an inefficient manner, the result will be less, not more, safety overall.

I do not mean to imply that States do not have an important role to play in pipeline safety. States now have safety authority over the approximately 1.5 million miles of intrastate natural gas lines, including the natural gas lines of local distribution companies. This is 75% of the total pipeline mileage so states already have jurisdiction over the vast majority of pipeline activity.

States also have jurisdiction over the single greatest opportunity to improve safety on interstate and intrastate systems: one-call systems that can help prevent accidents caused by parties digging into pipelines by mistake. Of the accidents that do occur on interstate natural gas pipelines, about 80% of those involving deaths or injuries are the result of these accidental dig-ins. Those most likely to be affected are the excavators themselves.

The industry has pleaded for stronger one-call programs for years and led the fight for a Federal statute aimed at encouraging states to strengthen their programs. Too often, state laws in this area exempt some excavators and don't provide for effective enforcement. While the cause of the Bellingham accident is still under investigation, it appears from what has been learned that excavator damage played some role in the accident. The Office of Pipeline Safety has sponsored an initiative called "Common Ground" and their June, 1999 report reviews the "best practices" found in one-call programs around the country. We urge the State to review its law in light of these recommendations and strengthen the law where appropriate.

• Public Right to Know

As I mentioned earlier, pipelines already provide a good deal of information to state and local organizations and to individuals who live along our pipeline rights of way. Still, we are willing to discuss what kind of additional information would be useful to local officials and residents.

I noted that Federal law already requires pipelines to have public education programs. We participate in one-call programs. We provide residents who live along our rights-of-way with information about the lines and phone numbers to call in the event of seeing anything unusual. We work with local emergency response officials to educate them about the nature of our operations and the appropriate actions to take if there is an accident. We encourage our employees to interact with local officials.

The advent of the Internet may provide another opportunity to educate interested persons in pipeline activities. The Office of Pipeline Safety maintains a web site with a wealth of information and it may be possible to build on that effort.

One additional word of caution. At times in the past, the industry has been warned by the federal government that pipeline facilities may be targeted by terrorists and asked to take steps to guard against possible attacks. For this reason, the industry has been wary of providing detailed information indiscriminately. Given the growth of the Internet, it is probably unrealistic to think this information could be kept confidential, but security concerns should be given some thought as part of this process.

CONCLUSION

If there is one point that I hope to impress upon you today, it is that safety is the result of a combination of activities that vary from company to company and location to location. Across the board mandates to conduct inspections at specific intervals or to conduct specific tests at specific intervals rarely make sense in practice.

In summary, the interstate natural gas pipelines are proud of our safety record. At the same time, we are actively seeking ways to improve upon that record, whether its by developing new and better technology, re-examining our current methods, or doing a better job of educating the public and excavators about pipeline safety. We certainly cannot promise that there will never be another accident, but we remain committed to taking every effective action we can to prevent the next one from ever occurring. We are also committed to working with the Office of Pipeline Safety as they do their job of overseeing the safety efforts of the industry. We hope that Congress will give us and the Office of Pipeline Safety the flexibility we need to accomplish this mission.

78-574 D-5

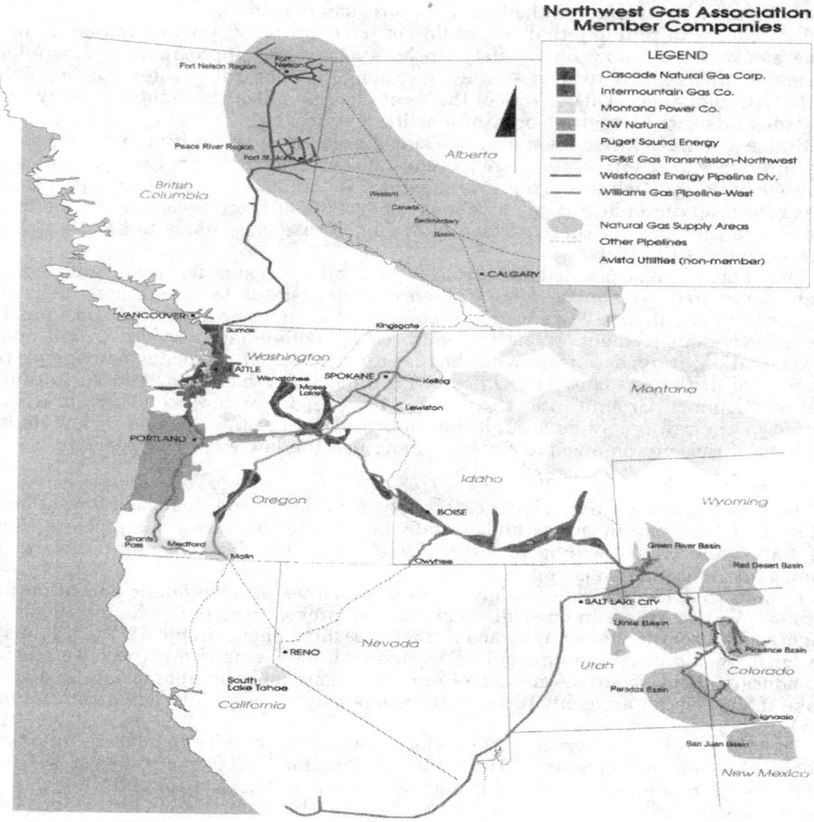
Northwest Gas Association
Member Companies
LEGEND
Cascade Natural Gas Corp.
Intermountain Gas Co.
Montana Power Co.
NW Natural
Puget Sound Energy
PG&E Gas Transmission-Northwest
Westcoast Energy Pipeline Div.
Williams Gas Pipeline-West
Natural Gas Supply Areas
Other Pipelines
Avista Utilities (non-member)
Fort Nelson Region
Fort Nelson
Peace River Region
Fort St. John
Alberta
British Columbia
Western Canada Sedimentary Basin
CALGARY
VANCOUVER
Sumas
Kingsgate
Washington
SEATTLE
Wenatchee
Moses Lake
SPOKANE
Kellogg
Lewiston
Montana
PORTLAND
Idaho
Oregon
BOISE
Wyoming
Green River Basin
Red Desert Basin
Grants Pass
Medford
Malin
Owyhee
SALT LAKE CITY
Uinta Basin
Piceance Basin
Nevada
RENO
Utah
Colorado
South Lake Tahoe
California
Paradox Basin
Ignacio
San Juan Basin
New Mexico

Transportation-Related Fatalities 1998

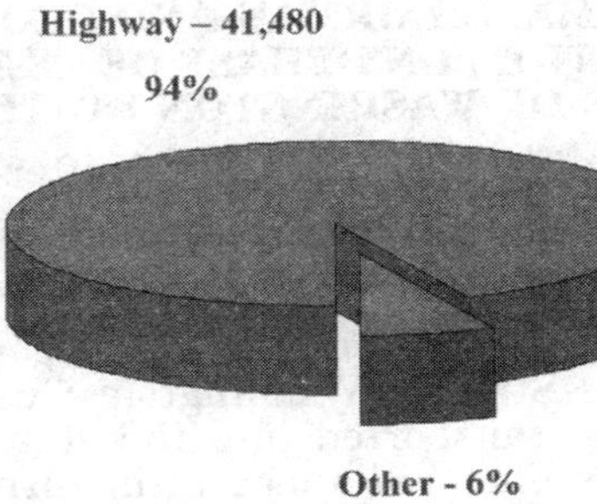

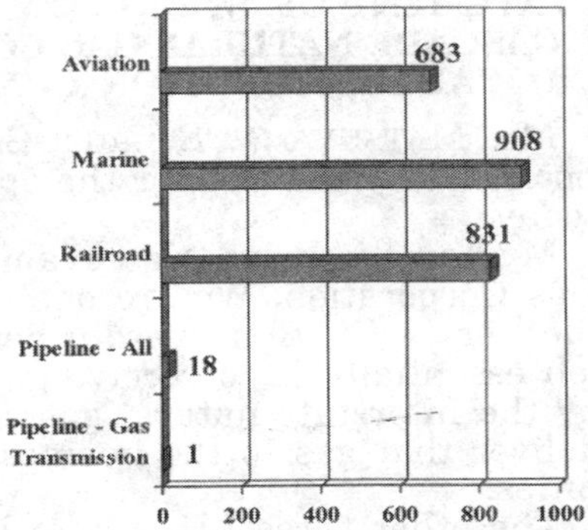

Sources: National Transportation Safety Board and the Office of Pipeline Safety, US DOT

Natural Gas Transmission Pipeline Accident Summary 1990 - 1999

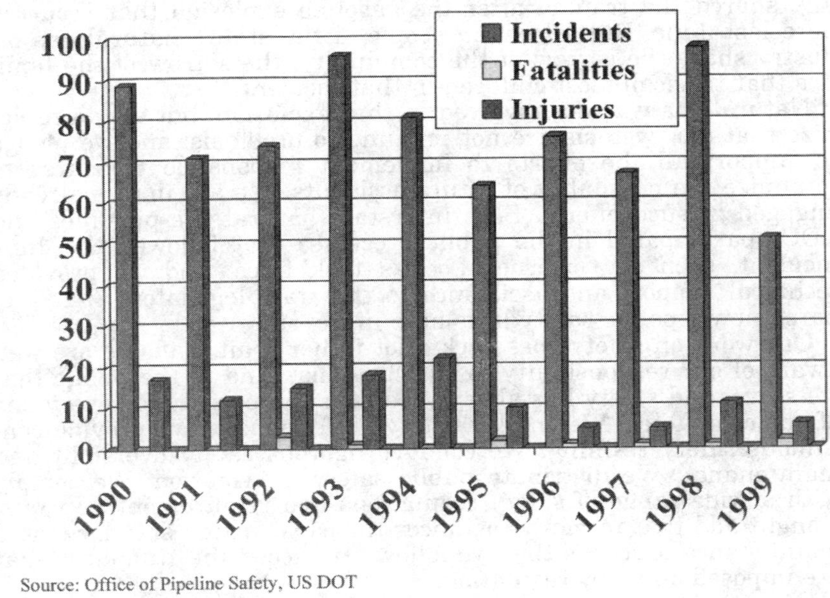

Source: Office of Pipeline Safety, US DOT

Senator GORTON. Through how many states do the pipelines run by your company pass?

Mr. SLUDER. In total all of Williams' gas pipelines, I think it's about 35 states, Senator.

Senator GORTON. Thank you.

Mr. Matsuyama.

STATEMENT OF W. BRIAN MATSUYAMA, CHAIRMAN AND CEO, CASCADE NATURAL GAS CORPORATION, ON BEHALF OF THE LOCAL DISTRIBUTION COMPANIES OF WASHINGTON STATE

Mr. MATSUYAMA. Senator Gorton, Senator Murray, good afternoon. I do thank you for the opportunity to be here and participate today.

My name is Brian Matsuyama. I'm Chairman of Cascade Natural Gas Corporation. We are one of the four local distribution companies' or LDCs' that operate here in the State of Washington. Our job essentially is to receive gas that is transported into the state by the interstate natural gas companies like Williams and in turn deliver that gas to the businesses and homes in our local communities.

The other three LDCs here in the state are NW Natural, Puget Sound Energy, and Avista Corporation. I'm appearing on behalf of all four companies.

My company, Cascade, is the company that serves Bellingham, and——

Senator GORTON. I think you might want to get a little closer to your microphone.

Mr. MATSUYAMA. My company does serve the City of Bellingham, and as members of the community many of our employees personally shared the trauma after the gasoline explosion that occurred here last June. But all of the people of the state's natural gas industry share the sorrow of the community, the sorrow of the families that tragically lost children in that incident.

Natural gas was not involved in that explosion, but we do recognize that our systems are not immune to problems; and we pledge to support all the efforts to implement responsible measures to minimize the possibility of future accidents. We are in fact already engaged in such efforts. Both interstate natural gas pipelines and LDCs participated in the public processes that followed the June incident, such as Governor Locke's task force, and we provided technical support and assistance to the state legislators that prepared new pipeline safety laws here in the state.

Our work on safety goes back a lot father than that. We are well aware of our responsibility to our customers and to the public that we serve, and safety has always been the top priority for every one of our companies. We employ safety professionals. We provide continuous safety training. We conduct rigorous safety oversight and maintenance. We distribute public safety information. We comply with a wide range of safety regulations and requirements. In fact, I might add that in many instances the construction standards and maintenance practices that we follow far exceed the standards that are imposed on us by regulation.

Our people are active in regional and national safety committees. They belong to numerous organizations dedicated to advancing safety. We support and contribute heavily to institutions that are engaged in developing advanced safety technologies. In 1999, the regional interstate natural gas pipelines and local distribution companies spent about 20 million dollars on non-construction, safety-specific activities.

Safety is a matter of corporate policy for every LDC. So we are, of course, extremely interested in the various measures that are being considered to improve public safety here in the State of Washington, and I'd like to summarize our perspective on some of those.

First of all, we take no position regarding the state regulation of hazardous liquid pipelines. I'd only note that we, the LDC's that we're talking about, are already subject to the jurisdiction of the state with respect to our intrastate operations, and that includes, of course safety matters. The Washington Utilities and Transportation Commission has long had jurisdiction over us in safety matters, and the staff has always done a conscientious and thorough job. Therefore, we do not at all question their ability to—given appropriate resources, and I think Ms. Showalter commented on that—we don't question their ability to expand the oversight of their responsibilities.

However, as customers of the interstate natural gas pipelines, we are dependent on them to provide safe, reliable, economic transportation of natural gas into our state.

In his testimony, Mr. Sluder identified some substantive issues regarding state regulation and jurisdiction over interstate natural gas operations. We LDCs, as their customers, simply urge the committee to give those issues due consideration.

We also urge the committee to avoid legislatively imposing specific prescriptive requirements for application both to liquid and natural gas pipelines. Our construction and operating characteristics and the materials that we transport are materially different. We are not at all suggesting that valid safety requirements be compromised. That's got to be number one, but we believe that regulatory processes, whether Federal or state can craft better requirements that fit particular operations than requirements from legislative processes. Failure to appropriately address industry differences for specific operating conditions could produce unintended consequences, and, of course, the worst of those consequences could be the reduction of safety rather than enhancement.

We are especially concerned about requirements that might impair the reliability of natural gas transportation service from the interstate pipelines, and thereby create new risks downstream of their systems.

Let me illustrate what I mean here. We are different from the liquid lines in that uninterrupted flows are critical to natural gas systems. If the interstate flows to us are interrupted, safety considerations require us to turn off service for each of our individual customers, then when flow is resumed, we have to go back and individually turn on every appliance in every home and every business that we serve. The reason that we do that is to ensure that raw gas isn't escaping from some pilot light. That process is long. That process is involved. That process obviously entails its own risk. I'm simply pointing out that a poorly designed requirement, no matter how good the intentions, could result in such loss of service to hundreds of thousands of customers just in Washington State, and that creates far more downstream risks than may be warranted by the upstream benefits.

I want to stress that we do not oppose measures to improve public safety. We merely recommend that they should be adopted only after careful study and with a clear understanding of all likely outcomes.

And finally, we support three specific recommendations that were made by the Governor's task force: the public's right to know and to understand how and where the natural gas system operates; the improvement of the one-call system; and advanced preparation and training for fire, police and emergency service personnel who are often the first to arrive at a hazardous site.

With that, I'd like to again thank you for the opportunity to share our views. We look forward to working with Federal, state and local authorities as well as within our industry to insure the highest degree of public safety.

[The prepared statement of Mr. Matsuyama follows:]

PREPARED STATEMENT OF W. BRIAN MATSUYAMA, CHAIRMAN AND CEO, CASCADE NATURAL GAS CORPORATION, ON BEHALF OF THE LOCAL DISTRIBUTION COMPANIES OF WASHINGTON STATE

INTRODUCTION

Good afternoon Mr. Chairman and members of the Committee. My name is Brian Matsuyama. I am Chair, President and CEO of Cascade Natural Gas Corporation. My company is one of four local distribution companies (LDCs) serving the State of Washington. I am pleased to have this opportunity to comment on the essential matter of public safety and our pipeline system on behalf of these companies.

THE PACIFIC NORTHWEST NATURAL GAS INDUSTRY

The natural gas industry, interstate and local distribution companies, provides a significant portion of the energy that drives the Pacific Northwest economy. In 1999, the natural gas industry delivered approximately 40 percent of the total energy used by 1.8 million residential, commercial, industrial and electric generation customers in the states of Idaho, Oregon and Washington. We operate 7,000 miles of pipeline within the state and employ nearly 10,000 of the state's citizens.

There are two interstate natural gas pipeline companies serving the state: PG&E Gas Transmission—Northwest and Williams Gas Pipeline—West. There are four local distribution companies in Washington State regulated by the Washington Utilities and Transportation Commission: Cascade Natural Gas Corporation, NW Natural, Puget Sound Energy and Avista Corporation.

THE NATURAL GAS INDUSTRY AND THE BELLINGHAM TRAGEDY

Let me begin by saying that my company, Cascade Natural Gas, serves the Bellingham area. This makes the tragedy that occurred in that community very personal. The people of the region's natural gas industry feel deeply for the families who lost their children in the Bellingham explosion on June 10, 1999. As neighbors, friends and fellow citizens, we share their grief and the trauma suffered by the entire community. We pledge our support to find and implement responsible steps that minimize the possibility that such an event could occur on the natural gas system.

Although the Bellingham incident did not occur on the natural gas system, we deemed it essential to take part in the assessment that followed. Governor Locke's formation of the Fuel Accident Prevention Response Team provided a forum to evaluate all areas of pipeline safety. Both interstate pipelines and local distribution companies followed and actively participated in the public process. We have remained involved by offering support, technical advice and comment regarding pipeline safety laws being prepared in the Washington State legislature.

NATURAL GAS INDUSTRY COMMITMENT TO SAFETY

The natural gas industry understands that its systems are not immune to problems. We also know that the public, our customers, depends on us to ensure their safety. The rapid growth of our customer base reflects the important role natural gas plays in providing affordable, environmentally clean and safe fuel. We are doing everything possible to sustain this confidence.

A recent AGA survey revealed that almost 55 percent of all American households used natural gas. In 1998, 69 percent of all new single-family homes were connected to natural gas. During that same time period, 64,000 homeowners switched to natural gas from oil, electricity or propane. In the Pacific Northwest, more than 90 percent of all new homes are connecting to natural gas when it is available. In order to retain the public confidence that underlies this growth, we recognize that the safety of our operations is critical.

Consequently, our industry has a tremendous incentive to maintain an enviable safety record. Safety is a matter of corporate policy and a top priority for every company. These policies are carried out in specific and characteristic ways: Each company employs safety professionals, provides on-going employee safety training, conducts rigorous system oversight and maintenance, distributes public safety information, and complies with a wide range of safety regulations and requirements.

Individual company efforts are supplemented by collaborative activities in the safety committees of regional and national trade organizations. Examples of these groups include the Pacific Coast Gas Association, American Gas Association, Interstate Natural Gas Association of America and the American Public Gas Association. Other safety-focused activities exist within industry-government groups. A current example is the Government-Industry Consensus Team (GICT); a specialized group formed to develop and maintain a voluntary data collection process that supports the analysis of plastic pipe characteristics. The GICT is composed of representatives from the American Gas Association, American Public Gas Association, the U.S. Department of Transportation's Research and Special Programs Administration (RSPA), the National Association of Regulatory Utility Commissioners, and the Plastic Pipe Institute.

Company safety professionals also participate in a variety of professional organizations dedicated to advancing the practice of work place and public safety. A partial list of the leading groups include the following: National Association of Corrosion Engineers (NACE), National Fire Protection Association (NFPA), National Safety Council (NSC), American Petroleum Institute (API), American Welding Society (AWS), American Society of Mechanical Engineers (ASME), American Conference of Governmental Industrial Hygienists, Transportation Safety Institute (TSI), American Society of Civil Engineers (ASCE), and the American Society of Safety Engineers (ASSE).

Companies also contribute to research and development through such organizations as the Gas Research Institute and Institute of Gas Technology where advanced safety devices and technologies are designed and tested. It is estimated that in 1999, the region's interstate pipeline and local distribution companies invested approximately $19.7 million in non-construction safety-specific activities. These voluntary expenditures reflect the high level of corporate commitment to public and employee safety.

AN LDC VIEW OF CHANGES IN WASHINGTON STATE PIPELINE SAFETY

Arguments are being made for state jurisdiction over safety regulation of interstate hazardous liquid and natural gas pipelines. The State of Washington, through the Washington Utilities and Transportation Commission, already has regulatory authority over LDC's intrastate operations, including safety matters. Our safety record attests to the conscientious, thorough approach taken by the Commission and its staff. Therefore, we do not question the ability of that agency, given appropriate resources, to oversee safety matters relating to the interstate pipelines, and we take no position regarding safety jurisdiction over hazardous liquid pipelines. However, we are aware of practical, substantive issues raised by the interstate natural gas pipelines regarding the imposition of a patchwork of state-by-state regulations. As customers of the interstate pipelines, and dependent upon them to provide safe, reliable, economic transportation of gas into our state, we urge the Committee to give due consideration to the concerns they raise.

We also urge the Committee to avoid legislatively imposing specific prescriptive requirements intended to be uniformly applied across both liquid and natural gas pipeline systems whose construction and operating characteristics are materially different. All of Washington's LDC's are deeply committed to safety, and we are not suggesting that valid operating requirements be compromised. However, we believe that regulatory processes, whether federal or state, are better equipped to craft requirements that fits the particular operation being regulated than are legislative processes. Failure to adequately address industry differences could produce unintended consequences, not the least of which would be to reduce rather than enhance the level of public safety.

We are especially concerned about requirements that might impair the reliability of natural gas transportation service from the interstate pipelines. Unlike liquids lines, uninterrupted flow is critical in natural gas lines. If interstate flows are interrupted, the ability of LDC's to maintain adequate pipeline pressures to serve end use customers is immediately impaired. In such situations we must manually turn off service to individual customers, one by one. When flows resume, we must then restore service and re-light each gas appliance in every affected home and business. The process is a long and tedious one, and is obviously not without its own risks. A poorly conceived requirement designed to improve pipeline safety, could result in such a loss of service to thousands of Washington customers, creating more downstream risks than are warranted by upstream benefits.

We support the public's right to know and understand how and where the natural gas system operates. An informed public will be better able to contribute to accomplishing the objectives of improved public safety. In many instances, improving public information will be a cooperative effort by the natural gas industry and communities served. Whether such efforts extend or improve existing programs or are new initiatives, local distribution companies will participate in their development and implementation.

At the present time, states have jurisdiction over the single greatest opportunity to improve safety on interstate and intrastate systems—the one-call system. We encourage and support the strongest possible one-call system for the State of Washington, including regulations for effective enforcement and graduated penalties for non-compliance. While the cause of the Bellingham incident is still under investigation by the Department of Transportation, it appears that excavator damage may have contributed to the explosion. The Office of Pipeline Safety has sponsored an initiative called "Common Ground." Their June 1999 report describes one-call "best practices" found in programs across the country. We encourage the State of Washington to review these recommendations when attempting to strengthen the one-call operation.

Local distribution companies will continue to support advanced preparation and training for fire, police and emergency service personnel who are often first to arrive at a hazardous site. It is critical for them to know and understand the nature of a natural gas incident and how best to manage it.

SUMMARY & CONCLUSION

The natural gas industry is an important component of the Northwest economy. The fuel the industry provides increases energy diversity, helps the environment, and is less expensive than most other energy options. It is also demonstrably safe. Nevertheless, the tragedy in Bellingham is a dramatic reminder that safety is the result of constant vigilance and a continuing effort to improve.

The natural gas industry is proud of its safety record. Natural gas has become the recognized fuel of choice throughout the Northwest. This customer growth and confidence also bears with it an added responsibility. As such, public and employee safety is the top priority for every natural gas company.

The natural gas industry believes that specific steps can and should be taken to improve public safety. However, these actions should be made only after careful study and with a clear understanding of all likely outcomes. Some recommendations of the Washington Governor's Fuel Accident Prevention and Response Team have significant potential—improved public information, an enhanced state one-call system, and training for emergency response personnel.

Thank you for providing the opportunity to present our views on the important matter of pipeline safety. We look forward to working with federal, state and local authorities, as well as within our industry, to achieve the highest possible level of public and employee safety.

Senator GORTON. Mr. Matsuyama, is Cascade entirely engaged in intrastate commerce exclusively regulated by the state utility commission?

Mr. MATSUYAMA. Yes. Senator. We serve two states. We also serve the State of Oregon, but we do not have activities that go across the state lines, in terms of transporting gas or lines that run between the two states so that our operations are entirely state regulated.

Senator GORTON. That's interesting. Is the regulatory regime significantly different from the regulatory regime in Washington?

Mr. MATSUYAMA. In the terms of the quality of the commission?

Senator GORTON. Uh-huh.

Mr. MATSUYAMA. It is not. They've been just as active in pursuing safety matters in that state.

Senator GORTON. Mr. Sluder, Williams on the other hand with respect to its operations in the State of Washington is exclusive interstate and governed by the Office of Pipeline Safety?

Mr. SLUDER. Correct.

Senator GORTON. So you have nothing to do with the state Utilities and Transportation Commission?

Mr. SLUDER. That's correct.

Senator GORTON. But I take it, and Mr. Matsuyama was fairly precise on this, but I take it that both of you believe that both the safety considerations, the potential for disaster with respect to natural gas transmission is different than that of liquid fuel and that they should not be subject to precisely the same kinds of requirements?

Mr. SLUDER. I'll take a shot at that first, Brian.

Yes, that is correct, Senator. There are various, there are many characteristics that are different. It was mentioned by one of the previous testifiers about the pooling of liquid. Our material does not do that. It is lighter than air, and so it escapes vertically. It does not pool or buildup in any concentrations on the ground.

The valving situations, valves are, are a better control in a liquids pipeline and a more viable, useful control than they are perhaps in a gas pipeline. Our first line of defense is to keep the gas in the pipe, to keep the pipe, itself, integral, be it from corrosion or dig-ins or landslide or whatever. So there are, there are numerous differences between the liquids and the gas substances and the way the pipelines react.

Senator GORTON. Senator Murray at the very beginning of the hearing showed us a chart indicating the number of fatalities from pipeline accidents over a period of, I think, it was 14 years. How many of those were from natural gas pipeline accidents? Do you have any idea, either of you?

Mr. SLUDER. I believe Senator Murray had a chart at one time that indicated that there, during that period of time there were 280 fatalities attributable to natural glass.

Senator GORTON. OK.

Mr. SLUDER. That is broken down by intrastate, interstate and then off-shore. All three of those components, each of those have different pieces of the 280 which you showed though the 14-year period.

Senator GORTON. Were more of them bystanders or employees?

Mr. SLUDER. I can give you the interstate piece of that. In that 14-year period, 13, I'm sorry, 39 of those fatalities were attributable to interstate systems. In 1 year, 22 alone were attributable to two off-shore platform accidents, where employees exclusively and contract personnel were the victims.

Senator GORTON. Thank you two for your testimony.

Now I'm going to turn to Mr. Gast.

Mr. Gast, Olympic is subject to a great deal of criticism here today. One of the statements that I found most remarkable because I hadn't heard it before was from the official from the state Department of Ecology who said that Olympic's safety record is far below industry norms, and when I specifically asked him does that mean that other interstate liquid, that pipelines operated far more safely than Olympic does, his answer was yes.

I'd like your—how is it that Olympic has so much worse a record than so many other liquid pipeline companies?

Mr. GAST. I assume that he was discussing leaks, those sorts of things is where he was coming from, as far as events.

Senator GORTON. Well, he was discussing the general reputation of Olympic as being relatively indifferent to safety and not meeting even minimal industry standards.

Mr. GAST. I don't, I have not made a comparison between Olympic and industry standards, so I really don't have those numbers.

Senator GORTON. Well, it might be well if you got them.

Mr. GAST. Senator, I certainly agree with that. I, right now at this present time, we're spending an awful lot of time with the corridor safety action plan and the other points that we're trying to do to reassure the communities and the public that we have a safe system here, and I'll guarantee you there are a lot of different areas that I need to go further into as we go down the road here.

Senator GORTON. Along that same line, all three of the cities south of Bellingham through which your pipeline passed spoke about the lack of cooperation on the part of Olympic when it sought information, and then the presentation of information in a form that an ordinary person or official simply couldn't understand, and at least one of them testified that rather than getting a map of the pipeline from the company, they had to hire outside consultants to look at the materials that you provide them and even to make that determination. Why that lack of cooperation with local officials?

Mr. GAST. Again, I can tell you what my experience has been since I've been here in this last couple of months is we have attended a lot of council meetings, city meetings, neighborhood meetings, and when I do talk to the folks that come back like the next day. I do hear the same thing is we weren't responsive. We didn't provide materials, and when I talk to my folks, I said, "Were you out there? Did you do these things?" They said, "Yes, we did."

I will be happy to sit down with the cities and talk out what the issues are. I think it's the same sort of thing. We need to get in a community of cooperation, talking with one another. That's what really needs to be developed here from what I heard today. I did hear a lot of angry people, and that upsets me. Believe me. It upsets me deeply. So I do want to get with them, sorry.

Senator GORTON. Again that rather pervasive anger, what is your answer to Mr. King's question why should under the circumstances Olympic be operating even south of Bayview?

Mr. GAST. The pipeline as it stands with everyone checking it out, and I'll tell you, you can imagine everyone is looking at this. I mean we have help from all over as far as reviewing the pipeline system, you know, from OPS, ourselves, from folks that we bring in from the outside as contract people to look at it, and basically what they see and what the pressure reduction that's been made, it is a safe system to operate according to all the technical experts in the field. Right now we're operating at 80 percent of where the normal operating pressure would be.

Senator GORTON. Well, I suppose I can't blame you for this, but it is rather frustrating to have at this hearing someone who's only been here 2 months from Olympic and who maybe naturally doesn't have answers to many of these questions.

I'm sure that both Senator Murray and I will have a number of written questions too, the answers to which you don't have here right now.

Obviously, the company wanted to create a new image here in Bellingham and the State of Washington, but it does mean it's rather difficult for us to get our questions answered.

Senator Murray?

Senator MURRAY. Well, Mr. Chairman, I share your frustration. You've obviously only been here a few months, and you have had the opportunity, Mr. Gast, to listen to the testimony today, and I think you would agree with us that the public confidence is not there.

Mr. GAST. Yes, I do.

Senator MURRAY. In terms of Olympic pipeline and reopening it, part of what I believe needs to be done at the Federal level is to pass legislation that I've introduced and Senator Gorton has co-sponsored through Congress to assure citizens that pipelines are inspected on a routine basis, that they have access to that information, that the pipeline operators have training and certification, that we develop testing that they feel is adequate and that that confidence is restored.

As we work this legislation through congress, hopefully soon, will your company be supporting or opposing that legislation?

Mr. GAST. The company, I'm sure, will support anything that drives the system to safer, safer pipeline operation.

Senator MURRAY. Will you be supporting the legislation directly?

Mr. GAST. When we are talking about the types of items that you mention, community involvement, emergency response, you bet.

Senator MURRAY. Will there be any portions of our legislation—I assume you've seen the legislation?

Mr. GAST. To be honest with you, I have not read through the whole thing. I have heard portions of it.

A VOICE. Then what are you doing here?

Senator MURRAY. Well, I would ask Mr. Gast if you could please have Olympic Pipe Line let us know directly in a public manner in response to this Committee if you are going to support the policies of the legislation that's being proposed, and again, I would just tell you, and obviously you're not going to answer, but this community needs to know that they have the confidence that Olympic and other pipelines are operating safely, and we want to move that forward.

I do have a couple of other questions for you. One of the concerns that we have heard from the local fire and police agencies is that part of the policies of Olympic do not include notifying police and fire immediately when problems are detected in the computer systems or in the operation system. Has Olympic pipeline changed its policy and procedure manuals at this point to assure that police and fire are notified immediately when a problem is detected?

Mr. GAST. When we have a problem, when they have a problem, such as a release or that sort of thing, absolutely, we dial the 911 number.

Senator MURRAY. But is it part of your policies and procedures?

Mr. GAST. Yes, it is. Yes, it is.

Senator MURRAY. To do that, my understanding at least at the time of June 10th accident and since that time that it's not part of the policy manuals. Has that been changed?

Mr. GAST. It's part of the policy manual that I know of since I've been there.

Senator MURRAY. OK. Since you've been there. Were you just hired by Olympic in January? Is that correct, or just this——

Mr. GAST. Yes, I actually transferred to Olympic Pipe Line in January.

Senator MURRAY. You worked for a different company before that?

Mr. GAST. Right.

Senator MURRAY. Mr. Gast, last Thursday, OPS told you that you cannot open the pipeline until further inspections are done. Can you tell us how Olympic will comply with that order?

Mr. GAST. Right now we have to respond to OPS's order and whether we have agree with what we're proposing and what the, versus what they have sent to us, and I suspect that we're not far from being at a point where we're, everything looks all right to them, that we would reopen the pipeline. It's a matter right now of just addressing issues of, to OPS and what they've sent to us.

Senator MURRAY. Can you tell me how you qualify your pipeline operators right now or what kind of identification requirements or testing requirements you have?

Mr. GAST. Yeah, we're actually starting along the program with the operator qualification requirements that the Federal Government is asking us to do by the year 2002, but we're starting the program, hopefully, we're done by the year 2001, but there are several things that are done with a pipeline operator. They have to go through what we call computer-based training. There are training modules that are on the computer. We actually put them, set them down with a simulator, simulating pipeline operations. We can induce simulated abnormalities in the pipeline and see how this, have them talk through it, what you they see, how it's doing. We put them through hydraulic courses. We talk about emergency resource issues, communications with the field. It's a whole myriad of——

Senator MURRAY. What is the length of education and training do you give to an employee before they're allowed to operate a pipeline?

Mr. GAST. It can vary with the individual, but minimum, we're talking several months before a trainee comes on board.

Senator MURRAY. What are the education requirements you have for those that operate the computers?

Mr. GAST. Basically high school education. Hopefully they will have some, might have some pipeline experience, and we can pull them in where they have operational experience in the field which is helpful, and in all cases you can't do that from the start, you know, if you don't have somebody that wants to come into the control center, so we go outside. We get them from other companies.

Senator MURRAY. Will your company oppose the national certification and training requirements that we have in our bill?

Mr. GAST. It's like the OQR?

Senator MURRAY. In our bill we will require national certification and training certificates.

Mr. GAST. I believe that falls right in line with what's being proposed with that.

Senator MURRAY. You mean by——

Mr. GAST. By actually the operator qualification requirements that are coming out of DOT.

Senator MURRAY. Mr. Gast, I have several other questions I will submit for the record,* since I don't know that you have the answers to them, but let me just say that I think it is absolutely imperative that Olympic work very hard to restore public confidence. It's unfortunate that someone isn't here who has worked there longer than yourself, and I don't blame you that being unable to answer those questions, because that is part of restoring the public confidence that I think is so important.

Thank you, Mr. Chairman.

Senator GORTON. One more question, Mr. Gast, that I think you can answer.

Will you support or oppose a delegation of authority from the Office of Pipeline Safety to the state and the Utilities and Transportation Commission, regulatory authority over the operations of Olympic in the State of Washington?

Mr. GAST. I would say what I would support would be a unified type of rule. I would hope that the state and the Federal Government could get together and come up with a uniformed set of rules that would apply. It would be great rules for all states in the United States. That would be my hope.

Senator GORTON. Well, that doesn't quite answer my question. Would you support or oppose a delegation of regulatory authority from the Federal Government to the state Utilities and Transportation Commission over your operations in the State of Washington?

Mr. GAST. I would prefer a single point of regulation.

Senator GORTON. —i.e. Federal?

Mr. GAST. The Federal Government.

Senator GORTON. Now that you've answered the question, OK. I think that we will thank this panel for its efforts here and go onto the last group, Mr. Weimer and Ms. Harper.

OK. This has been a long afternoon, and we'd like to get this last panel an opportunity to speak.

Mr. Weimer.

*The information referred to can be found in the Appendix.

STATEMENT OF CARL WEIMER, EXECUTIVE DIRECTOR, ON BEHALF OF SAFE BELLINGHAM, BELLINGHAM, WASHINGTON

Mr. WEIMER. Senators Gorton and Murray, thank you for this opportunity to testify. I appear on behalf of SAFE Bellingham, and with me is David Brickland, who has been active in public safety issues up and down the Puget Sound corridor, and he can answer technical questions, questions about franchise agreements or whatever you desire.

My father-in-law, now retired, was a tankerman for most of his career. He was responsible for loading and unloading fuel and oil barges from refineries and oil tankers throughout North America. He liked his job, and he raised a family that I'm now proud to be a part of. Like any family there are subjects that we avoid at family meetings. The dealings of the oil industry is one such subject. So I was surprised recently when he began the tell me why he thought that the Olympic pipeline should be reopened again. He told me of a recurrent nightmare that he still has after all these years. It is a fictitious event where he mistakenly overfills a barge and spills crude oil in to Puget Sound.

That nightmare along with his years of experience transporting oil over the water lead him to the conclusion that pipelines are the safer form of transportation. He went on to say that he feared that if the Olympic pipeline did not reopen, it was only a matter of time until some sort of terrible spill occurred in Puget Sound.

I kept my thoughts to myself. I didn't say that barge safety had changed substantially in the last 10 years or that the accident records really don't support his conclusions. I didn't even say that in reality we need all of the different modes of fuel transport so they should all be as safe as possible.

What I did come to realize was the terrible amount of responsibility and pressure that the oil industry has placed on its employees. Why does one of the richest industries ever created on the face of the earth continue to put profits ahead of safety? They could easily invest in the many high tech advancements that could now prevent many spills and help relieve their employees from this terrible worry.

If safety had been the highest concern then employees of Olympic Pipe Line would not now have to be taking the 5th Amendment. If safety had been the highest concern, then three families would not now have to grieve the loss of their children, and if safety had been the highest concern, then members of our community along with the employees of Olympic Pipe Line Company could sleep through the night without reoccurring nightmares.

Unfortunately, pipeline companies throughout the Nation have not put safety first. Even more difficult to understand, the government regulators who are supposed to keep pipelines safe also have failed to keep pipelines safe.

In the aftermath of our local tragedy we have learned many things, and sadly one of the first things we learned was that some of the most fundamental aspects of pipeline safety are not properly addressed in current Federal law. Consider the following: current Federal law requires pipelines to be pressure tested only once before operation can commence, never again. There's no Federal regulation requiring that internal inspection devices be used. There is no Federal requirement for pipeline operators to utilize leak detection systems. There's no federally defined Federal minimum standards for ensuring that a pipeline cannot be overpressurized to the point of breaking the pipe.

Further, Federal reporting requirements are inadequate to help us recognize current problems or protect us from repeating the same mistakes. Valve placement in current Federal law does not take into account the number of important factors in determining the adequate number and placement of valves to limit the volume of potential spills. And there are no Federal requirements for safety management in audit practices that Congress determined were needed to protect the public in other energy-related industries.

You will hear from the industry that it is in their own self interest to regulate themselves and avoid spills, but the sad truth is that in much of today's corporate world the focus is on the bottom line, not the bottom line 5 years from now, the bottom line this year, this month, today. Industry risk assessments may make sense to the corporate executive in Houston, but how do they explain to a family with a dead child that the profits are better than spending a tiny portion of those profits to ensure that a child's life is not lost. Congress should act quickly to mandate Federal safety standards before we lose another innocent life.

We are encouraged that OPS may delegate authority to the State of Washington, even so there are numerous examples where state regulatory agencies have also failed to protect human health and the environment. In the wake of the Exxon Valdez accident, Congress mandated the formation of an oversight council. We believe that for locally elected officials and citizens to once again have confidence in the pipeline companies serving our community, and the regulators who are supposed to oversee them, that a similar oversight council needs to be established.

We would hope that the new pipeline legislation or the reauthorization of the Office of Pipeline Safety, itself, would mandate the creation of a model local advisory council here in Puget Sound based on the Alaska model which was meant to be replicated.

The purpose of this council would be to review, monitor and comment on all fuel transportation operations, proposed operations and environmental impacts. The whole advisory council would play a major role in increasing public awareness of pipeline safety, spill response, spill protection and environmental protection issues. The local advisory council's members would be comprised of representatives of appropriate interest groups such as local municipalities and counties, tribes, environmental organizations, fishing organizations and agricultural.

This proposal is designed to promote partnership and cooperation between local citizens, industry and government. It will also go a long ways toward rebuilding the trust that was shattered when the pipeline exploded on June 10th, 1999. We believe that the local governments and citizens that have the most to lose if something goes wrong with the pipeline should have an active role in the oversight of this pipeline. A model local advisory council, established here in Puget Sound, which if successful could later be replicated in other parts of the Nation would serve as a crucial link for this local involvement.

In conclusion, if there's one message that I hope you take back to Washington, D.C. it is that this event is not unique to Bellingham. Pipeline spills occur almost daily. Similar spills and explosions have happened nationwide. According to the Office of Pipeline Safety in the past 15 years 342 people in 41 different states have been killed by pipeline accidents, which spilled over a hundred and thirty million gallons. The property damaged from these spills was over a billion dollars.

The time is now for pipeline regulations to be updated to include a whole range of safety improvements that are now possible. Pipeline companies need to be required to invest in safer pipelines, and state and local governments need to be allowed to set stricter pipeline regulations if they deem them necessary.

Finally, and perhaps more importantly, those most impacted if something goes wrong transporting fuel need to be included in all aspects of the transportation of fuel through their communities.

Referring to a different northwest problem, Senator Gorton recently summed up our view perfectly when he said "it's time for the Federal Government to let those who will be affected by the decisions help make those decisions."

Thank you for your time.

[The prepared statement of Mr. Weimer follows:]

PREPARED STATEMENT OF CARL WEIMER, EXECUTIVE DIRECTOR, ON BEHALF OF SAFE BELLINGHAM, BELLINGHAM, WASHINGTON

Senator Gorton, I appear today on behalf of SAFE Bellingham. Thank you for this opportunity to testify.

My father-in-law, who retired a number of years ago, was a tankerman for most of his career. He was responsible for loading and unloading fuel and oil from barges, refineries, and oil tankers from Alaska to Puget Sound to Puerto Rico. I believe he liked his job. He certainly made a good living, bought a fine home overlooking Fidalgo Bay, and raised a family that I am now proud to be a part of. My children are his grandchildren.

Like any family there are subjects that we avoid at family gatherings. The dealings of the oil industry is one such subject, so I was surprised at a recent birthday celebration when he began to tell me why he thought that the Olympic Pipe Line should be reopened again. I bit my tongue and listened as he went on to tell me of a recurring nightmare that he still has after all these years. The nightmare, which still regularly wakes him, is of a fictitious event where he mistakenly overfills a barge and spills crude oil into Puget Sound. That nightmare, along with his years of experience transporting oil over the water, led him to the conclusion that pipelines are safer than the forms of fuel transport he is knowledgeable about. He went on to say that he feared that if the Olympic Pipe Line did not reopen it was only a matter of time until some sort of terrible spill occurred in Puget Sound, perhaps right outside his window on Fidalgo Island.

I kept my thoughts to myself. I didn't say that barge safety had changed substantially in the past ten years, or that the accident records really don't support his conclusion. I didn't even say that in reality we need to use all the current modes of fuel transport, so they should all be as safe as possible. What I did come to realize was the terrible amount of responsibility and pressure that the oil industry has needlessly placed on its employees. Why does one of the richest industries ever created on the face of the earth continue to put profits ahead of safety? They could easily invest in the many high tech advancements that could now prevent many spills and help relieve their employees from the terrible worry, and associated nightmares, of responsibility for needless accidents?

If safety had been the highest concern than employees of Olympic Pipe Line Company would not now have to be taking the Fifth Amendment. If safety had been the highest concern then three families would not have to grieve the loss of their children. And if safety had been the highest concern then members of our community, along with the employees of the Olympic Pipe Line Company, could sleep through the night without recurring nightmares.

Unfortunately pipeline companies around the nation have not put safety first. Even more difficult to understand the government regulators who are supposed to keep pipelines safe also have failed to put safety first. That is why we are here today—from the Governor of the State of Washington to an ordinary citizen like myself—asking that you join with us to regain control of the laws that govern pipeline safety.

SAFE Bellingham is a community based organization which did not even exist on June 10, 1999—the day that the Olympic Pipe Line exploded in a fireball in Bellingham, Washington killing three young people and plunging a fiery dagger in the heart of this community. SAFE Bellingham came into existence shortly after this tragic event as the community tried to come to grips with the pain, shock, and sorrow. In the aftermath, we have learned many things. We have learned about weaknesses in the federal oversight of petroleum pipeline safety. We have learned about the federal government's efforts to prohibit state and local governments from protecting their citizens from tragedies like these. And we have learned that pipeline companies—driven by this year's bottom line—sometimes this quarter's bottom line—do not have the economic incentive to deal with hidden risks that may not explode onto the scene until some later year. We come here today to share our new found insights with you.

While our organization may be new, the information we bring to you is not. If there is one message that I hope you take back to Washington DC it is that this event was not unique to Bellingham Washington. Similar spills and explosions have happened nationwide. According to the Office of Pipeline Safety in the past fifteen years 342 people in 41 states have been killed by pipeline accidents which spilled over 130 million gallons. The property damage from these spills was over a billion dollars. The only really unique thing about the disaster here in Bellingham, thanks in large part to the Senators from Washington State, is that the U.S. Senate has come to town to consider ways to make sure that such a disaster never happens again. We hope the rest of the Senate will also remember the 337 other families, in 40 other states, who have not had the benefit of such a hearing.

I would like to focus on four things we have learned in the aftermath of the Bellingham accident: the inadequacy of current federal regulations; the inappropriate federal preemption of state and local safety regulations; the absence of adequate self-regulation by Olympic Pipe Line Company; and the need for effective community involvement in overseeing pipeline safety issues.

INADEQUACY OF CURRENT FEDERAL REGULATIONS

Sadly, one of the first things we learned was that the petroleum pipeline industry has escaped the safety regulations that have made so much of America a safer place in the last part of the 20th century. Some of the most fundamental aspects of pipeline safety are not addressed or are addressed inadequately in current federal law. Consider the following:

Testing and Inspection of Pipelines

Current federal law requires pipelines to be hydrostatically pressure tested only once—before operations commence. These pipes then sit in the ground for years and decades. They are subjected to corrosion and strain from earth movement. They are subjected to enormously high operating pressures. But there is no requirement that these pipelines ever be hydrostatically pressure tested again.

Pipelines can also be inspected internally by the use of so-called "smart pigs." These internal inspection devices use different technologies such as ultrasonic or magnetic waves, to try to detect some (not all) anomalies in the pipe. Different types of pigs have different strengths (and weaknesses) in detecting different types of anomalies. Over the last decade or two, some vendors have developed smart pigs with much better capabilities than earlier versions. While not perfect, these devices provide a useful tool for determining the condition of the pipe after it is put into service. Ironically, while federal regulations require that new pipelines be designed to accommodate smart pigs, there is no federal regulation requiring that the smart pigs be used.

There also is no oversight of how pipeline operators use (or neglect) the information they receive from these internal inspection devices. In Bellingham, Olympic had run a smart pig through the pipe five years before the accident. The inspection device had detected numerous anomalies but Olympic chose to ignore most of them. Only after the tragedy was Olympic forced (by the Office of Pipeline Safety) to dig up the pipe in numerous additional locations where anomalies had been detected and finally determine whether the pipe's integrity had been compromised. Notably, before Olympic undertook a remedial hydrostatic pressure test this fall, it first repaired a dozen or more sections of the pipeline that had been identified as containing anomalies five years earlier.

Leak Detection

There is no federal requirement for pipeline operators to utilize leak detection systems. If a computer based leak detection system is used, there are only very general regulations specifying how it should be configured. These standards fall far short of assuring reliability. Thus, the Bellingham explosion was preceded for more than an hour by a huge rupture, which dumped almost 300,000 gallons from the pipeline. There was so much gasoline flowing down Whatcom Creek that it turned the creek into a river of gasoline. Yet Olympic's controllers 100 miles away in Renton, Washington apparently were unaware. For more than an hour, the gasoline gushed with no warning to the controllers because of the unreliable leak detection system.

Management Audits

Even a properly designed and constructed pipeline becomes a menace to the communities through which it crosses if it is not operated and maintained well. If settings on safety valves can be adjusted in the field with no comprehensive oversight, there is a problem. If safety procedures are not updated when new facilities are added to the line, there is a problem. If operators cannot find records of what type of pipe they have in the ground, there is a problem.

Other segments of the petroleum industry have been required to adopt extensive safety management practices and undergo safety management audits. See, e.g., 29 C.F.R. § 1910.119. But these federal requirements have not yet been made applicable to petroleum pipelines. In fact, the pipeline industry has been specifically exempted from the very safety management practices Congress determined were needed to protect employees and the general public in other energy related industries.

Right-of-Way Encroachment

Current regulations require right-of-way minimums of 50 feet—but this is waived if there is at least 12 inches of dirt covering the pipe. This makes no sense. The industry is aware that construction backhoes and other heavy equipment are a major source of damage to pipelines and such equipment has the capability of reaching far more than 12 inches below the ground surface.

There also is no regulation assuring that a pipeline operator responds appropriately once notified of construction in the right-of-way. In Bellingham, for instance, there are indications that in 1994 Olympic knew heavy construction equipment was operating in very close proximity to its pipe (where it ruptured in 1999) but Olympic did not provide continuous oversight of the construction to prevent damage to its pipe.

Avoiding Over-Pressurization

There is no federally defined minimum standard for assuring that a pipeline is "fail-safe," *i.e.,* that it cannot be over-pressured to the point of breaking the pipe. Many pipeline companies install pumps that can create pressures greater than the pipes can withstand. They then use pressure safety valves to prevent over-pressurization. But there are no federal regulations assuring that these critical devices are adequately designed, installed, maintained—or even used! In the Bellingham explosion, for instance, there are indications that a key pressure safety relief valve was improperly selected and/or installed. The NTSB report suggests that it operated numerous times in the preceding months—far too frequently for a valve that was supposed to activate for emergencies only.

Further, without adequate oversight, there is a danger that a supposedly redundant backup safety system will be just as susceptible to failure as the primary system. For instance, in the Bellingham accident, flawed input data apparently caused Olympic's main computer to fail. But then when the backup computer came on line, it was fed the same flawed data and it failed, too. There is also no redundancy when both a primary and backup system rely on the same power source. With no regulatory standards established, the industry is free to make critical mistakes like these.

In the end, the best protection against over-pressurization is assuring that the pumps are not sized so large that they can create pressure to the point of bursting the pipe. But there is no federal prohibition on over-sizing pumps.

Valve Location Requirements

The strategic placement of check valves and block valves can do much to limit the volume of a spill. The very general, minimum requirements in current federal law do not address spill volume, elevation changes, and other factors such as terrain and population density that must be taken into account in determining an adequate number and placement of valves.

Inadequate Reporting

We can learn from the past only if we know what has happened in the past. Current federal law seems designed to frustrate our ability to learn from past mistakes. The duty to report to OPS certain "safety conditions" is waived if the condition is corrected within a particular timeframe. Spills of less than 2,100 gallons generally need not be reported. The City of Bellingham recently negotiated with Olympic Pipe Line an agreement that requires Olympic to report these "safety conditions" regardless of whether they are corrected within the timeframe and virtually all spills regardless of size. The federal government should establish similar reporting requirements.

FEDERAL PREEMPTION

Ironically, at the same time that federal law is so inadequate in mandating safety requirements for the pipeline industry, federal law simultaneously prohibits state and local governments from adopting their own safety-based standards. Congress is well aware that state and local governments can regulate safety and environmental protection without compromising industry's ability to operate on a national and international level. To take but one obvious example, the trucking industry, whose fleets criss-cross our state borders thousands of times a day, are subject to safety requirements at the state and local level. *See* 49 U.S.C. § 14501(c)(2). As long as state and local safety regulations do not interfere with interstate commerce and do not conflict with federal requirements, Congress has seen fit to allow this additional level of protection.

Yet when it comes to oil pipelines, Congress has set a different standard. Here, Congress prohibits state and local government from protecting their own citizens from the calamities that can befall them from a poorly designed, operated, and/or managed pipeline. The sooner states and local government are given the power to protect their citizens, the sooner we will see significant advances made in safety protection for this industry.[1]

THE PIPELINE INDUSTRY DOES NOT ADEQUATELY REGULATE ITSELF

You will hear from the industry that it is in their own self-interest to regulate themselves and avoid spills of valuable petroleum products and explosions, which cause the deaths of innocent people. But the sad truth is that in much of today's corporate world the focus is on the bottom line—not the bottom line five or ten years from now, but the bottom line this year, this quarter, this month—TODAY. Short-term financial incentives frequently push managers in the direction of maximizing revenues and profits at the expense of risks, which may not manifest themselves for years or even decades. Industry risk assessments may make sense to the corporate executive in Houston, but how do they explain to a family with a dead child that profits are a better bet than spending even a tiny portion of those profits to ensure that a child's life is not lost?

This approach is not unique to the oil pipeline industry. As a society, we have responded to this phenomenon by imposing safety standards on other industries. Those standards are far and above those that have been required of the pipeline industry. How many more tragedies likeBellingham's must be witnessed before we see the same level of safety mandated for petroleum pipelines as we do for other segments of the petroleum industry? Hopefully none. Congress should act speedily to mandate federal safety standards before we lose a single more innocent life.

COMMUNITY OVERSIGHT

In the wake of the Exxon Valdez accident, Congress (in the Oil Pollution Act of 1990 (OPA 90)) mandated the creation of a well-funded, independent, non-profit citizen oversight council. *See* 33 U.S. Code, § 2732. We believe that for local elected officials, and citizens to once again have confidence in the pipeline companies serv-

---

[1] This is not the first time that the states have called for stronger federal regulation of petroleum pipelines and a relaxation of federal preemption. Following the pipeline explosion and fire in Moundsview, Minnesota, the Minnesota Commission on Pipeline Safety investigated and deliberated over four months and concluded that federal regulations of petroleum safety had to be increased and that federal preemption of state regulations should be relaxed. These recommendations were finalized in December 1986, nearly 13 years ago and circulated widely in Congress. Sadly, the federal government has failed to fully pursue, adopt, and implement most of the Commission's recommendations. Perhaps Liam Wood, Wade King, and Stephen Tsiorvas would be alive today if it had.

ing our communities, and the regulators who are supposed to oversee them, that a similar oversight council needs to be established. We would hope that any new pipeline legislation, or the reauthorization of the Office of Pipeline Safety itself, would mandate the creation of a model citizen council here in Puget Sound, based on the OPA 90 model which was meant to be replicated. The purpose of this council would be to review, monitor, and comment on pipeline companies' risk management and risk assessment studies; their spill response and prevention plans; their prevention and response capabilities; their safety and environmental protection programs; and their actual impacts on the environment.

The Citizen Advisory Council would play a major role in increasing public awareness of pipeline safety, spill response, spill prevention, and environmental protection issues. The Citizen Advisory Council would have no law enforcement or regulatory authority but would have the same access to pipeline facilities and records as state and federal regulatory agencies. Like the OPA 90 language, the proposed legislation should direct federal agencies to cooperate with and consult with the Citizen Advisory Council on substantive matters related to pipeline operations.

The Citizen Advisory Council's members would be comprised of representatives of appropriate interest groups such as local municipalities and counties, tribes, environmental organizations, fishing organizations, and agricultural groups.

This proposal is designed to promote partnership and cooperation between local citizens, industry, and government. It would also go a long way towards rebuilding the trust that was shattered when the pipeline exploded on June 10th 1999. We believe that the local governments and citizens that have the most to lose if something goes wrong with a pipeline, should have an active role in the oversight of those pipelines. A model Citizen Advisory Council established here in Puget Sound, which if successful could later be replicated in other parts of the nation, would serve as the crucial link for local involvement.

In conclusion please remember that the accident here in Bellingham was not unique. Pipeline spills occur almost daily. Pipeline regulations need to be updated to include a whole range of safety improvements that are now possible. Pipeline companies need to be required to invest in safer pipelines, and state and local governments need to be allowed to set stricter pipeline regulations when they deem them necessary. Finally, and perhaps most importantly, those most impacted if something goes wrong with a pipeline need to be included in all aspects of the pipelines running through their communities. Referring to a different northwest problem Senator Gorton recently summed up our view regarding pipelines perfectly when he said "It's time for the federal government to let those who will be affected by the decisions make these decisions."

You have our support, and we wish you luck, on the pipeline reform measures you are supporting. We hope you will include our proposals as well. Thank you for your time today.

Senator GORTON. Thank you very much, Mr. Weimer.

Miss Harper.

STATEMENT OF SUSAN HARPER, EXECUTIVE DIRECTOR, CASCADE COLUMBIA ALLIANCE, SEATTLE, WASHINGTON

Miss HARPER. Thank you. Thank you for holding this hearing, Senator Gorton and Senator Murray.

My name is Susan Harper. I'm Executive Director of Cascade Columbia Alliance. I feel privileged to be here to speak on behalf of the coalition of citizens, elected citizens, environmental civic groups, labor and businesses that I'm representing with my testimony.

I'm also on the board member of the National Pipeline Reform Coalition to promote pipeline safety nationwide.

The people who originally came together as the NPRC did so for the reason that we are here today. Elsewhere at another time before Bellingham was a tragedy from a pipeline accident that equaled in magnitude the grief and confusion that we feel in this room today.

Through our association with the National Coalition since 1996 when the Olympic Pipe Line proposed a 230-mile Cross Cascade pipeline, we learned many of the lessons that could have prevented a Bellingham nightmare.

We urge the Committee to take swift action, to set right a tragedy that should have never happened by listening to the recommendations for safety improvements that CCA and others have presented today.

I would also the request that this packet of about a hundred letters from concerned citizens be included in the record.*

Senator GORTON. It will be without objection.

Miss HARPER. Thank you. Make no mistake that this is not an isolated incident. Broadly in terms of pipeline safety nationally, but also the company that manages Olympic Pipe Line is called Equilon, and most likely the gentlemen speaking today was an employee of Equilon. Equilon also runs a refinery at Anacortes, Washington that exploded last November 1998, killing six workers. That means within 7 months Equilon's safety record includes nine fatalities.

The Pipeline Safety Act in 1996 was actually weakened and moved toward a risk management demonstration project-based procedure. We're listening to testimony today that gives evidence to this lack of, of regulation and this moving to pipelines being self-regulated, and in fact, Jim Hall's testimony in 1997, excuse me, in 1998, or 1999, sorry about that. It's a little nerve racking.

Senator MURRAY. Miss Harper, I should tell you, she's getting married this weekend.

Miss HARPER. I am, and I'm just a little nervous, but he did testify in 1999 before the house that there were 28 recommendations before the Office of Pipeline Safety, and they followed up on those in 1997 only to find up that seven of those had actually been implemented.

CCA wholly supports the bills that are before Congress right now, and we're very appreciative of those. I want to emphasize the importance for an independent oversight body to monitor and review the plans of the industries and the agencies to protect our interests, and you've heard today about this regional citizens oversight committee in Prince William Sound. In fact, Alaska Governor Knowles credits the regional oversight body as creating the safest oil transport system in the United States.

The State of Washington is clearly in need of such a model, and we would request that part of the reauthorization of the Pipeline Safety Act includes such an oversight model as we have been discussing here today, and I know Senator Gorton has supported in his talking points.

We have compiled a long list of safety recommendations that we've attached to our testimony.* However, we do not agree that pipelines are necessarily the safest means of transportation, but we would agree that all are inherently dangerous and that we need to take steps to make sure that all of those systems are made safer including double wall tank ships, double wall trucks and double wall pipelines where needed.

---

* The information referred to was not available at the time this hearing went to press.

Congress must all require that the industry carry out the recommendations of the NTSB and that OPS make that, or remove any blockage of doing that.

During our review of Olympic's proposed Cross Cascade pipeline, here I have a copy of the draft EIS. It's kind of small, actually, you'll note for such a large pipeline. During that time, we were very concerned about the I–5 pipeline, and in fact, we requested that Olympic's existing I–5 pipeline be inspected and replaced as an alternative since the pipeline was built in 1965 and had 44 spills on record with OPS. Our coalition was very concerned about the condition of the pipeline that we were being told was oversubscribed. Not only did Olympic refuse, but Jones and Stokes, the consultants hired to prepare the draft environmental impact statement to this pipeline, were not directed to or decided not to disclose information about the pipeline's condition. Those who had full access to information concerning Olympic's existing woefully inadequate and unsafe pipeline system continuously reassured the public that this pipeline was safe and that Olympic's leak detection system was adequate.

In the draft EIS Jones and Stokes gave high marks to Olympic's existing supervisory control data and acquisition system, and assured the public that major leaks could be detected and shut off within minutes. This false and rosey picture was completely blown away by Olympic's devastating June 10th explosion.

In our view those who had access though this information bear responsibility for misleadingly the public about the risks from Olympic's existing pipeline.

As a society we also need to reduce our reliance on this non-renewable energy. We need to enforce automotive fuel economy standards, and we appreciate Senator Gorton's support for CAFE standards.

In addition, for an independent citizen oversight body that can address safety issues, the Pacific Northwest is solely lacking in a basic energies fuels planning strategy similar to the Northwest Power Planning Act.

We would also in hearing the testimony today, I would like to bring up one additional thing that wasn't mentioned in here. I know that Secretary Slater has written a letter saying that the UTC could be the best place to have delegated authority for pipeline safety. During the Cross Cascade pipeline which was proposed to be an intrastate pipeline, the UTC was silent. They decided not to give comments on this pipeline proposal. However, the Department of Ecology was very involved.

Also, the UTC during their rule adoption period when they got to the delegation of intrastate regulation for liquid fuel pipelines, we were participating in that process, and unfortunately all of our recommendations to go beyond the Federal standards and all of our evidence that was presented at that meeting was basically ignored, and the Federal code were adopted, and the UTC settled on the Cross Cascade pipeline saying that they would defer to the Federal codes.

As you can see, we have real concerns with the agency that will be overseeing pipeline safety in Washington State. However, we know that any agency does need to have good follow-through and oversight over it. That's why we would so wholly support the citizen monitoring organization that we've been referred to in Alaska called the Regional Citizens Advisory Council. We'd also wholly urge that we need to raise the floor for pipeline safety laws nationally in order to make sure that we are having a process that is coordinated through one agency that has adequate standards to do its job which is to protect the public interest and the environment, and the last thing that I would like to say in conclusion is that we would request your full support in getting legislation passed this year.

We are planning a national pipeline safety conference on April 9th and 10th in Washington, D.C. because we are very aware that this needs to be a national issue and that the State of Washington needs the support of the entire nation to make this happen. We're very excited to let you know that we've got about half the states with activists in them that are willing to come to this or public officials or pipeline safety professionals, and we would urge you to support that fully, and it would be a really great way to actually be able to make these changes, and finally, the Colonial Pipeline, I have the record for that pipeline, by the way. It's the largest, or it's a very large pipeline, and it has one of the worst spill records in the U.S. Eight point five million gallons have been spilled since 1968 with a total of a 186 spills, and 60 of those were in the last 5 years, and the fact sheet that we created was created in early 1999. So as you can see the Office of Pipeline Safety needs some help from us, and we're willing to give it.

So we'll really appreciate that not only do we get delegated authority but we get citizen oversight and thank you for your time today.

Senator GORTON. Thank you. Perhaps both of you can comment on this, but Mr. Weimer, you first, how does the citizens' oversight council differ from the peer review councils that exist under the present act?

Mr. WEIMER. My understanding is the way the citizen oversight panel was set up in Alaska, it's totally independent, and it's actually funded by the oil companies up there, but there's a contract between the oil companies and the citizen oversight that make it independent, and it's designated in the Federal legislation instead of being within the department, like underneath the Office of Pipeline Safety. It's an independent functioning body with funding of its own that hires experts to look at different proposals of the pipeline companies or their environmental records.

Senator GORTON. That's the nature of your recommendation?

Mr. WEIMER. That's our nature. We'd like something that's independent, and there's already examples even within Washington state, and there's already a group that has formed, a city-county forum, and number of the cities' mayors were here today, and they've been meeting, and such a group I think could be turned into an oversight panel fairly easily, and that way they would be spending the oil companies' money hiring experts to try to figure out what pig runs mean instead of having to spend their own taxpayer moneys to do that.

Senator GORTON. Miss Harper, you agree totally with that?

Miss HARPER. Yeah, I would agree with that, but I would add that the current advisory body that sits under OPS, it's reactionary. The people who sit on that body are chosen by OPS, and in fact, Lois Epstein, one of the engineers from the environmental defense fund, has been very chagrined at the recent selection and one of our own, Greg Winter, was somebody that we put up to be on that committee. He served on the regional citizens advisory council in Alaska as a chair of one of those committees, so we feel that he was very well capable of doing the job, but unfortunately his access was denied. So there isn't a process that is set up that you have to go through to get on that committee.

Now, with the regional citizen advisory council, because first off the money is mandated, this council does not have to pay homage to the industry, because they're going to get their money no matter what.

Secondly, as Carl had mentioned, the pipeline safety forum, I've been an extremely active, or I've been a very active part of that. In fact, I've been on one of the steering committees, and I find that working with local governments as we're doing is a very good way to work, and I found that environmental, tribal, fishing, recreational groups and local governments really make a good mesh, because you're not, you know, just leaning too fair in one direction. You're really working with pragmatic people who are looking out for public safety, but also water quality protection, pipeline spills underground. In fact, in Delaware there was just a recent spill that we've heard has been going on for 12 years from a city plant there, and those kind of spills are just unacceptable and with water resources so scarce today, we really need to be careful with those resources.

Senator GORTON. Senator Murray?

Senator MURRAY. Thank you, Mr. Chairman, and thank you for your testimony and your support of our legislation, and we look forward to working with you.

It's been a long day, and it's I just have ask one question. In order to get this passed at the national level, the awareness has to be extremely high. Here in our state, a lot of people are involved in local communities up and down the I–5 corridor and even in eastern Washington. How are you working with your national organization to insure that that awareness is just as high, and do you find that more and more increasing awareness or is it still a challenge?

Mr. WEIMER. I think there's a tremendous amount of awareness. There's certainly been media coverage of this. There was a 500,000 gallon spill Friday in Texas. Texas has a fairly bad record of pipeline spills. There was a spill in Delaware. There was a spill in Philadelphia not long ago. So there's a whole lot of reasons for people being active. I spent all weekend on the phone with a mayor from New Jersey, a mayor from New York, a property rights organization in Pennsylvania helping to organize this national conference in early April, which we're trying to bring together fairly rapidly which is somewhat insane, but the support is there, and people are going to be there in Washington, D.C., to talk about pipeline issue.

Miss HARPER. In fact, we've all the been down to Texas to speak to a group down there. Pipeline proposals are also a big puzzlement for people, and in fact, this was an Ashland Marathon project that was proposed that the project was just pulled off the table because very same pipeline had a major rupture.

In Ohio, we have people. In Illinois we have people. In Montana, you know about the Flathead Indian reservation and their issues there. We have support in Alaska. So we definitely have a web of support, and I think that, the tragedy that happened in Bellingham can be the type, most unfortunately I say this, but also optimistically I say this, the Exxon Valdez of the pipelines.

We can't accept anymore deaths, Senators. I want you next time there's a tragedy like this that happens to anybody, any community to be able to say we did everything we could to prevent this. We're sorry, and if you can go that far then we're on the right track.

Senator MURRAY. Thank you very much.

Senator GORTON. Thank you. That's quite appropriate end to this day, Miss Harper. Thank you.

Miss HARPER. Thank you.

Senator GORTON. Remember that anyone who wishes to comment has 10 days to submit written comments to the Commerce Committee in Washington, D.C., and those comments will be included in the record.

With that, we're adjourned.

[Whereupon, at 6:30 p.m., the hearing was adjourned.]

APPENDIX

Written Questions Submitted by Hon. Patty Murray to Carl Gast

1. Mr. Gast (Olympic)—Last Thursday, OPS told you that you couldn't reopen the pipeline until further inspections are done. Could you explain how you intend to comply with that order?
2. My first question is for all of you. Pre-Bellingham, which most of us feel created an incentive for all of you to do a better job—certainly too costly of an incentive—what types of inspection and testing procedures did you use? How often did you inspect?
3. How do you qualify your pipeline operators? Are there education requirements?
4. Do you have any initial concerns about the integrity rule that will be up for comment soon?
5. Do you have any thoughts or concerns on how we could establish a "right to know" standard that would require you to establish a dialogue with those whom your lines affect?
6. What types of R&D activities are you involved with? Are you interested and do you think it would help for the federal government to increase investments in R&D?

[Response was not available at the time this hearing went to press.]

Joint Prepared Statement of State Senator Harriet A. Spanel, 40th District and State Representative Kelli Linville, 42nd District, Washington

Thank you to Senators Gorton and Murray and to the Senate Commerce Committee staff and Chairman John McCain for holding this important hearing in Bellingham. Your commitment to addressing this critical issue is greatly appreciated. In our combined 21 years of public service, there has been no greater tragedy to befall the citizens we represent.

Speaking as a senator and representative in the Washington State Legislature, we address you as fellow elected officials—those of us who try to make government the best it can be—and whose primary purpose therefore is to represent the interests of the citizens we hear from on a daily basis. Pipeline accidents happen here, in the cities and towns and counties we inhabit and which we are elected to represent.

As such, the question that keeps jumping out at us over and over again is "How could this happen?"

We have read the federal Pipeline Safety Act. On first reading, we are encouraged: so many aspects of pipeline safety are addressed, so many regulations are to be developed, so many apparent assurances that things will be ok.

But, like the pipelines themselves, this is a false hope of safety. It has not gone as possibly the authors of the federal act had hoped. For whatever reason, the guidelines of the federal act—which preempt anything we can do here at the state level—have not been implemented. Our constituents' anger is easy to understand.

And while the Office of Pipeline Safety dawdles, accidents continue to happen all over the country. Just since the first of this year, in Clark County, Kentucky, a pipeline burst and spilled 900,000 gallons of crude oil into a tributary of the Kentucky River—possibly the worst spill in the state's history. Drinking water was being shipped in, the smell was awful, and cattle watering holes were polluted.

A week later, in Toms River, New Jersey, lack of leak detection was blamed for a spill of 56,000 gallons of crude oil into the John Heinz Wildlife Refuge near Philadelphia. A visitor to the refuge, not the company, detected the spill. The pipeline in question was 50 years old.

And, committee members may remember that earlier this year Koch Industries was fined an unprecedented $30 million on account of its "wait to see if it breaks" leak detection system.

We have attached a short list of some significant pipeline accidents in recent years. Please note that *none* of these accidents were the result of "third party damage" with the exception of the Reston incident. *All* of them could have been prevented—if NTSB safety recommendations had been acted upon.

The common causes of pipeline accidents are:

- anomalies in the pipe not detected or not acted upon
- operator inattention or error
- computer system malfunction
- shut-off capability insufficient or improperly deployed
- leak detection insufficient

Because of the sorry record of federal regulation, we heartily endorse Senate Bill 2004 which Senator Murray introduced and Senator Gorton has sponsored. S. 2004 not only expands state authority to allow higher standards for training and leak detection, but it also pushes the Office of Pipeline Safety to enter into further agreements with states for inspection of interstate lines. The bill also provides much needed funding for the federal effort.

In anticipation of enhanced federal regulations and a stronger federal-state partnership for pipeline safety, we worked diligently in the State Legislature this year to obtain passage of House Bill 2420. This bill sets the stage for the increased authority S. 2004 allows. It also strengthens Washington's programs for preventing third-party damage and responding to accidents.

But we all know that the state law alone will not prevent further accidents. That is up to the federal government. And there is simply no reason to wait any longer for meaningful steps to be taken—to wait for more deaths and environmental disasters.

It would be presumptuous of us to reiterate the extent of the impact on the dear families who have lost children. You have heard their grief But we do not believe it is presumptuous of us to tell the federal government that we are not going to sit idly by and let a similar tragedy befall yet another community. Accordingly, we pledge to work with you to accomplish the maximum improvements possible in pipeline safety regulation.

Thank you, again, for your consideration.

Examples of Major Pipeline Accidents (1980–1999)

Accidents

(1) *Fredericksburg, VA* 1980 (and again in 1989)

330,000 gallons of aviation fuel entered the city water supply, and the Rappahannock River, shut down the water treatment plant, a state of emergency was declared, and businesses and residents hauled water for a week.

Causes: Pipe damage upon installation, subsequent undetected corrosion, operator error, insufficient valve placement.

(2) *Moundsview, MN* 7/8/86

An 8-inch gasoline pipeline burst, gasoline flowed along neighborhood streets until it was ignited, killing 2 people who burned to death and injuring 7.

Causes: Failure to correct known defects, inadequate pipe specifications, inadequate operator training including delay in responding.

(3) *Flathead Indian Reservation, MT* 1986–1993

Seventy-one leaks and three major spills of gasoline, aircraft fuel, and diesel (including 163,000 gallons into a creek) over this period resulted in the Flatheads refusing to renew Yellowstone's franchise and move the line off of the reservation.

Causes: Inattention and failure to correct defects.

(4) *North Blenheim, NY* 3/13/90

A liquid natural gas pipeline burst sending 100,000 gallons of product flowing down into the town—enough to engulf the entire town. Residents noticed a "heavy fog" on their windshields, until one called and notified a company employee. Two people were killed and seven injured.

Causes: Negligent maintenance procedures resulting in cracks in the pipe which were undetected; operator error; insufficient remotely operated valves and check valves.

(5) *Reston (Herndon), VA* 3/28/93

Pipe burst sending a geyser of diesel fuel (407,000 gallons) into the storm sewer and eventually into a tributary of the Potomac River. (Could have been gasoline or jet fuel.) Significant environmental damage ($1 million clean-up).

Causes: Third-party damage causing corrosion which went undetected for a long period.

(6) *Edison, NJ* 3/23/94

Natural gas transmission line burst and exploded. 1500 residents evacuated and $25 million damage. Injuries included minor burns and cuts from broken glass.

Causes: Line hadn't been "pigged" since 1986, but it had deteriorated; no remote automatic valves; pipe manufacturing standards lax; no extra measures for highly populated areas.

(7) *Allentown, PA* 6/9/94

Natural gas pipe burst and product flowed underground into the basement of an 8-story retirement home, where it migrated through vents into other floors and was eventually ignited. One death and 55 injuries.

Causes: Company employee (backhoe operator) error; no excess flow valves which had been recommended by NTSB since 1972; insufficient public awareness.

(8) *Reedy River, SC* 6/26/96

Fuel oil pipeline crossing the river burst resulting in a $20 million clean-up effort.

Causes: Pipeline corrosion not responded to soon enough; computer malfunction; employee error; inadequate leak detection.

(9) *Lively, TX* 8/8/96

Liquid natural gas pipeline burst, killing two men who accidentally ignited it.

Causes: Inadequate corrosion protection.

(10) *Murphreesboro, TN* 11/5/96

84,000 gallons of diesel fuel (could have been gasoline) and $5.7 million damage.

Causes: Corrosion; operator error—3½ hours before detection.

(11) *San Juan, PR* 11/21/96

Thirty-three people killed when a liquid natural gas line exploded.

Causes: Employee negligence in responding to a leak which had been ongoing for a week.

(12) *Bellingham, WA* 6/10/99

Prepared Statement of James Terrence Montonye, Technical Program Director for the SPIE, International Society of Optical Engineering

SPIE, which is located just south of Western Washington University in Bellingham at 20th and Knox streets, is the largest international professional society serving developers and appliers of optical, optoelectronic, and imaging instrumentation. It's imaging conferences range from x-ray to radar to acoustical. It is foremost in imaging spectroscopy and deals with the diagnostics of materials of all kinds.

SPIE employs 145 people and has gross revenues of $20M. It produces proceedings for 350 nominal two-day topical conferences per year and does about 350 short courses for engineers per year at twenty-four different meetings around the world. It has 13,500 members.

It happens that SPIE holds a non-destructive testing meeting in California each March that has produced three proceedings titled "Nondestructive Evaluation of Utilities and Pipelines." These proceedings contain papers by instrument developers from universities, laboratories, OEM corporations, and non-destructive testing contractors. Although we've tried hard to get them there, state and municipal engineers seldom attend this meeting.

Here's my thought. The nondestructive testing meeting in San Diego is a meeting where state pipeline inspectors could meet each year to work on realistic standards for pipeline inspection in the presence of people who develop the instruments and contractors who use them. Materials and materials testing will continue to evolve. Wealth creation through the conversion of natural materials into consumable products is a never ending process. Similarly, standards in important areas such as pipeline safety should be recognized and organized for a continual process of evolution, as well. I cannot think of a better way for the evolution of efficient and effective pipeline safety standards to occur than at a meeting of technical experts on the instrumentation required to do the testing.

Any electrical, materials, or mechanical engineer should be able to handle the job of inspection and standards development. But, there is another very important consideration in regulation enforcement. Jack London brought it out in "Fish Patrol." His effectiveness as an enforcement official he attributed to his youth as an oyster pirate. This same principle holds for pipeline inspectors. There are two major hazards with any federal or state inspection organization, graft and incompetence. We all are aware of many examples of both. The best way for states to handle this problem is to pay high wages to pipeline inspectors who had been engineers within the pipeline industry in other states. The high wage and out-of-state experience would serve to dispel graft. Engineering experience from within the industry would serve to provide the competence to get the job done efficiently and effectively.

Senator Gorton, these comments are respectively offered for your consideration following this public hearing on pipeline safety in the City of Bellingham on 13 March, 2000.

○